Reoccupy Earth

Reoccupy Earth

gROUNDWORKS|

ECOLOGICAL ISSUES IN PHILOSOPHY AND THEOLOGY

Forrest Clingerman and Brian Treanor, *series editors*

Series board:

Harvey Jacobs	Catherine Keller	Norman Wirzba
Richard Kearney	Mark Wallace	David Wood

Reoccupy Earth

Notes toward an Other Beginning

David Wood

Fordham University Press | *New York 2019*

Fordham University Press has no responsibility for the persistence or accuracy of URLs for external or third-party Internet websites referred to in this publication and does not guarantee that any content on such websites is, or will remain, accurate or appropriate.

Fordham University Press also publishes its books in a variety of electronic formats. Some content that appears in print may not be available in electronic books.

Visit us online at www.fordhampress.com.

Library of Congress Cataloging-in-Publication Data available online at https://catalog.loc.gov.

Printed in the United States of America
21 20 19   5 4 3 2 1
First edition

Contents

Contents

Introduction:
Reinhabiting the Earth

Something strange is the soul on earth.
                        —*Trakl*[1]
Understanding is like knowing how to go on.
                        —*Wittgenstein*[2]

Today's big news stories—the wars, the eco-disasters—all seem to
have the same gaping hole in them. This hole is lack of awareness,
and its thrum, once you begin to hear it, soon becomes deafening:
We can't go on like this.
                        —*Robert C. Koehler*[3]

What counts is the question, of what is a body capable?
                        —*Deleuze*[4]

Life on Earth

I have a lawyer friend who believes that it will help save the planet
if we turn off our engines while idling at traffic lights. The unnecessary CO_2 generated in this way could push us over the edge. I do
not subscribe to this view, but it captures, writ small as it were, a
widely held view that I do accept—that if we as a species are headed
for disaster, the streetcar we are traveling in is named Habit. This is
both true and important. But quite what is meant by "habit" here,

and what the truth of this claim requires of us, as citizens and as philosophers, is less clear. That is the question I explore here.

By way of orientation, let me explain where I am coming from. I recently spent five years running an interdisciplinary faculty seminar on climate change with a historian, a theologian, an anthropologist, a social scientist, a physicist, and a lawyer. Our hermeneutic sophistication could not get around the fact that climate science was painting an exceedingly troubling picture of the future. We swiftly moved to trying to understand why we continue toward the rapids with our foot on the gas pedal. We stopped laughing at lemmings!

At the same time, I have been teaching the later Heidegger—*Contributions*, "Letter on Humanism," and his writings on language and poetry.[5] He insists that when he speaks of dwelling poetically on earth, he is not talking about the housing shortage, or anything quite so practical, but of what one would normally think of as an existential reorientation.

Prima facie this looks like a standoff between the reality-based concern to save our species and the spiritual concern to save our soul. I argue that these two concerns can, in fact, be productively brought together, guided by the logic of the Möbius strip, where absolute opposition at any particular point bleeds into productive continuity. The possibility of this convergence has deeper philosophical implications. And the point d'appui of my reflection is habit.

It would be hard to overestimate the role of habit in our lives. At one level, this is all well and good. There are, of course, bad habits, which we try to kick, but our daily life would collapse without the scaffolding of habit. Still, when we contemplate climate change and the catastrophic future it portends, it is hard not to conclude that "we cannot go on like this." Business as usual simply cannot continue for long.

By "business as usual," I mean the common cloth of our Western daily lives, our normal practices, in large part consisting of habits—personal, collective, economic, and intellectual. A contemporary green Socrates would be buttonholing people on the street, asking them whether what they were doing was really sustainable. The difficult thing to deal with here is that while most of our habits are, on a certain scale, under certain conditions, perfectly reasonable, when aggregated they may spell disaster. Moreover, our habits reflect and are integral to narratives of the good life, social norms and expectations, as well as economic realities. Our commitment to them is not necessarily to their specific shape and contour but to the fact that these are the ones we have, and that some such shared narratives and norms are necessary for a full life.

If business as usual is tied up with the shape of our habits, and if this shape reflects a more basic attunement or form of life, which may be shared globally whether as a fact, an aspiration, or a destiny, and if business as usual cannot continue, we have a problem. We may conclude that this basic attunement rests on hardwired facts about human nature or that even if one could imagine another way of life, we cannot get there from here, given people's current preferences. This path leads to despair, or resignation. If there is nothing to be done, well then. . . . We might, for example, come to believe that the earth cannot support its projected human population at current western standards of living, and that, sadly, the problem will be "solved" by war, disease, famine, or new forms of gross inequality. Many will conclude that the rational solution is for each of us to anticipate this prospect—individually, collectively (racially, nationally, culturally)—and shore up our own prospects with wealth, walls, and property. Even guns.

Forms of life, patterns of dwelling, other than our current consumerist model are clearly possible. Human history, with its archive of alternatives, is our teacher here. But whether we can get there from here voluntarily, and who precisely "we" are, is another matter.

I take philosophy to be a multifaceted practice of emancipation, whether from illusions, myths, injustice, or oppression. It achieves these ends indirectly by showing us that what we think of as natural and necessary is not—that other paths may be possible. We can call this "critique" both in the sense of laying bare the operations of power and in the Kantian sense of exploring less obvious conditions of possibility for meaning, truth, and experience. It is not by philosophy alone that we may reinhabit the earth. But if reinhabiting means changing some of our deep habits, habits reflecting historical sedimentations and congealings (hopes, fears, promises of happiness, practical necessity), then unearthing the forces in play, seeing how they operate and what is at stake in reconfiguring them, is a historical task to which philosophy can at least contribute. Philosophy is not alone here. Economists are central to imagining other economic orders, such as that of degrowth. Psychologists and social scientists can teach us (up to a point) what really makes people happy. Political scientists and historians can draw lessons from earlier revolutions and other dramatic changes (conquests, migrations, technological innovation).

In Part I, we open by outlining the promise of a couple of new kids on the block: eco-phenomenology and eco-deconstruction. Other philosophical approaches—analytic, pragmatist, critical theoretical—have complementary virtues; my own approach reflects in no small measure the traditions that have most influenced me. When Husserl launched phenomenology with his "back to the things themselves," the *Sache* he referred to was the vital stuff of experience. Going *back* to it reanimates concepts that would otherwise be lodged in a state of stranded reification. In principle, an ecologically oriented phenomenology would experientially counteract the toxic effects of hubristic conceptualization through which we often connect with the natural world. I argue for a phenomenology that does not seek refuge in being simply descriptive, as Husserl sought, but takes seriously its capacity for edification, for renewal. Later, in Heidegger's hands, phenomenology more strongly opens up alternative ways of thinking about our earthly dwelling, ones that point to a reinhabiting of the earth.

Deconstruction, while not directly a friend of the earth, charges us with critically examining all manner of exclusion, violence to the other, and symbolically (textually) mediated forms of silent subordination. While it has no time for naive naturalism, it supplies powerful tools for exposing the metaphysical humanism on which our apparent planetary dominion is ideologically grounded. Pursuing what I have dubbed eco-deconstruction, I graft an eleventh "plague" onto Derrida's remarkably down-to-earth list of the ten plagues of the New World Order (in his *Specters of Marx*). This shows, once again, that deconstruction is not some hermetic hermeneutic or arcane textual practice but fully attuned to the textures of the real.

Finally, Part I harnesses Whitehead's process philosophy to the service of an ecological imagination. His wide range of conceptual innovations advances an environmental and contextual understanding of human existence, specifically encouraging what I have called a temporal phronesis, articulating the full range of our modes of temporal engagement with the world. This is especially productive when thinking about urgency, prediction, uncertainty, irreversible change, long-term horizons, and so on, all of which cut into simplistic linear models of progress and all of which have a particular relevance when thinking about climate change. The ecological imagination is not about visions of paradise but about developing alternative shapes of experience, critical ways of reading, and, as Nietzsche might put it, *dancing* with our spatial and temporal hermeneutic practices.

Part II opens up something of a second Copernican revolution with respect to the masterful subject. It explores a panoply of experiences that disrupt our habitual existence, that challenge our everyday complacency, by reversal, transformation, and estrangement. In "Things at the Edge of the World," I argue for what I call a fractal ontology, one in which what we think of as the furniture of the world either has the capacity to project a world of its own or can be seen to be constitutive in some important way of the world, of which it initially appears merely to be a part. The sun, for example, is both a large item up there in the sky and the material-transcendental condition of our very existence, indeed of life itself. Through these experiences, we find ourselves to be vulnerable and receptive, dependent in ways that are not obvious. While many of these examples focus on "individual" experience, one whole dimension of transformation opens us up to the depth of our constitutive connection to and dependence on other humans, and other creatures, both through "consciousness-raising" experiences of solidarity and through active group participation.

"Touched by Touching" treads just this line. Carnal sensuousness, including but not limited to sexuality, is a tacit dimension of everyday seeing, smelling, hearing, and touching, but it also arrests us, brings us up short, effecting a break with humdrum practicality, reminding us, in something of a reversal of Plato's anamnesis, of the delights of embodiment, inciting us to be "true to the earth."

Part III opens onto a wider horizon, with reflections on the deep relationship between embodiment, place, and territory, which lay the groundwork for both contestation (territory) and sharing (the earth). In "My Place in the Sun," the contestation of place and territory through historical narrative confirms the essential temporality of place, not least when claims to property are being made, even as it leaves open the fate of claims that make such appeals. It highlights the problematic status of the body in its primitive occupation of space on a finite planet. Our embodiment grounds both violence and the necessity of negotiating a shared civil space with others who, as Kant argues, equally have a "right to be somewhere."

Time returns in "On Being Haunted by the Future": the future is another site of contestation. If deconstruction from the beginning broke with a purely linear time, Derrida later asks us to contemplate a to-come, an im-possible future, a messianism stripped of theology. While this is set against a purely calculative time (prediction through extrapolation and induction), we argue that there is another irresponsibility, flowing into culpable negligence, that must not be ignored—when an all-too-predictable disaster is allowed to happen. We must anticipate the predictable as well as expect the unexpected.

Part III concludes with an unfashionable (but spirited!) defense of an enlightened anthropocentrism, one aligned with a certain biocentrism from which it is claimed to be inseparable. Breaking again with habitual thinking, we claim that it is in our own interest not to understand our self-interest as opposed to that of our fellow travelers on the planet. Moreover, if what makes humans special is our commitment to values such as truth, justice, and compassion, we surely would have to consider willing our own demise as a species if we came to believe that our continuation threatened these values. (Imagine a transposed version of Rawls's veil of ignorance scenario.) And we can only claim Reason as our unique gift if we demonstrate our capacity to deploy it by collective sustainable agency.

But more needs to be said about the philosophical stakes here, in connecting habit with inhabiting, and then the possibility of re-inhabiting (the earth).

Philosophy as Dehabituation

Philosophy is the enemy of habit. And habit of philosophy. Even a philosopher who championed habit would have had to step back from the "taken for granted" in order to do so. Philosophy thrives on wonder, on doubt, on critique, on questioning—and each time it is not just "common sense" that suffers but the dispositional structures we call habits that bear the brunt of the scrutiny.[6]

We can provisionally distinguish six different modes of dehabituation:

1. **Wonder**. This captures both puzzlement and perplexity, not to mention awe. We don't understand how or why things are the way they are. Or we are astonished that anything exists at all. Clearly here, habit doesn't cut it.

2. **Angst**. Why am I here? Who am I? What should I do? This can open up new possibilities—or be paralyzing.

3. **Questioning**. I have Socrates in mind here. We use words like "justice," "love," "time," and "truth," but do we really understand what we mean by them?

4. **Critique**. Stepping back—analyzing the way things are, either with a view to change them or to try to understand why they are or must be just so.

5. **Poetic/spiritual dehabituation**. We come to see that some third thing, or fundamental consideration, is at issue, albeit silently, in our ordinary relation to the world. Our understanding of Being, the Other, God, and transcendence are all contenders here.

6. **Deconstruction**. (Building on 4 and 5.) Deconstruction "makes tremble" the pursuit of the value of "presence" in a text.

The habit that is "dehabituated" is not a particular empirical habit, such as biting ones nails or eating cornflakes for breakfast, but a certain assured naive unreflective disposition, which may in some respects be admirable or delightful. No one is saying that the child gleefully building a sandcastle on the beach should from the outset be reminded that the tide will later wash it away. When Foucault speaks of the very idea of man in just terms, we nod appreciatively. But, again, we might want to set aside our hesitations when supporting Amnesty's campaigns for human rights.

These different modes can clearly be further differentiated, and supplemented. They may also function as preparatory stages for one another. We may begin to ask Socratically about the meaning of justice, and then, having got a bit clearer, move on to critique, perhaps even protest and revolution, to try to bring it about.

That is not so say that philosophy is always victorious. Many philosophers have left their desks endorsing or championing the significance of habit in one way or another. Think of Hume and backgammon after dinner.

Philosophical Habit

But philosophy has not itself been immune from habit. Many philosophers have charged their fellow thinkers, even the history of philosophy, with having got into unreflective ruts, a repetition that, in effect, reproduces within philosophy the very condition it seeks to root out elsewhere. This is true of Kant, Nietzsche, Heidegger, and Derrida, for example. I call these habits, hopefully not stretching the term too far, because if we are to believe the critics, they are unconscious, repeated, and, importantly, avoidable. As we might also say, they are cognitive, textual practices reflecting a certain desire.

It might be said that (for Kant) applying experiential categories beyond their proper scope (a tendency of the mind?) should not be called a habit. And do I really want to claim that status for ressentiment? Or for Seinsvergessenheit, the forgetfulness of being? Or, for that matter, one-dimensional thinking, "metaphysics," misplaced concreteness? Some responses: first—of course—I am not (yet) endorsing the claim that these habits are the problems they are said to be. Second, I see no reason not to include higher-order, cognitive dispositions under the heading of habits. Third, it seems to me entirely to be expected that claiming something to be a habit should at times be a matter of contestation. My test is: is it (or is it claimed to be) a common, regular, largely unconscious disposition to which we can at least imagine alternatives? The word "unconscious" here does mask a problem. I do not claim there is a sharp line dividing conscious and unconscious. We may be aware that there is "something wrong," or feel unease in a certain situation, without being fully able to articulate the source of the problem. We just may not have the new concepts or language that would be needed to capture the experience. Think of a static essentialist in a dynamic world. Or a scientistic mind faced with beauty. Nonetheless, I want to defend the idea that many of the ways in which we commonly think should be seen as habits. And we may learn something from Kant's sense (and Wittgenstein's) that things go "wrong" when we deploy these dispositions beyond their proper scope of application. The fact that we may indeed do this would be further evidence for their being habits precisely because we do not notice our crossing the line. One thinks here of linear thinking, projective thinking, wishful thinking, calculative thinking, and, yes, meditative thinking. Each can be or become a "habit" when deployed indiscriminately.

It is important, too, to add that many a decision taken "fully consciously," as we might say, could be said to be made under the influence of habit. More of this later.

A Critique of Habit

What is true individually is true collectively, too: without habits, we would not make it through to tomorrow. So much of who we are, how we think, what we eat, when we do things, where we spend our time is the direct outcome of habit. As the scaffolding on top of our biological instincts, our drives, and not always clearly to be distinguished from them, habits infuse our being-in-the-world and coordinate our social existence. Our capacity to reflect may make us distinctive as living creatures, but our ability to learn to do things without thinking is the essential foundation on which it all rests. Obviously, particular habits do not always serve us well, but, here again, having the second-order disposition to correct the application of first-order habits when needed, again without thinking, is essential, as when we swerve to avoid a dog on the road or, to take a clearer example, when we steer into a skid to avoid losing control of a vehicle. Habits are constitutive of our most complex behaviors. Some are taught, but all are learned. Habits can be taught habitually, without thinking. Spanking a misbehaving child was a normal, handed-down piece of behavior, as was the justification for it.

But when, in connection with our terrestrial existence, it is said that "we cannot go on like this," it is our habits—again both individual and collective—that are the problem. Some of our habits, or the particular form that they take, will have to be changed. And yet we are so heavily invested in many of them that we do not even see them as habits, and our willingness to change at the required speed is missing. While it may not be necessary or sufficient to enable the kind of change we need, I would like to sketch what I would call a critique of habit in the double sense of critique (both exposing conditions of possibility and imagining alternatives).

Let us just list some of our terrestrially toxic behaviors to start with, without deciding ahead of time which are properly called habits:

Emission of greenhouse gases
Pollution of sinks, such as atmosphere and ocean

Destruction of nonhuman life-forms (Sixth Great Extinction)
Population increase
An economy of growth

These five categories are each dangerous in terms of their consequences for the global ecosystem when they reach a certain scale.
Many are themselves consequences of other behaviors rather than
conscious commitments. We do not set out to generate CO_2, or
methane. And in terms of the scale of the problem, there is no "we"
to address, anyway. We may deliberately dump sewage and toxic
waste into rivers and the sea. But no one sets out to "pollute" the
ocean. We have long imagined it was self-cleaning, like a tidal beach.
And if you pee into your local stream on a summer afternoon, that
is still true. With rare exceptions (smallpox, bedbugs, wolves), we do
not set out to eliminate other species. But territorial encroachment,
the use of pesticides, and climate change have that consequence. As
for population increase, well, as I understand it, most babies, even
when welcomed, were not "intended." No one could be said to be
responsible for the human population explosion. The odd man out
in my list might be thought to be the economy of growth, linked
to consumerism, to free trade, to a refusal to acknowledge the constraints set by natural capital, all backed by impressive economic
theory. For, by and large, it is explicitly advocated and promoted,
often linked, not implausibly, to other shared goals such as reducing
unemployment and promoting social justice. Some of its negative
consequences—threatening local economies, political disempow-
erment, new concentrations of wealth—may be accepted as unfortunate and predictable but inevitable. Even here, the fact that the
economy of growth encourages the externalization of costs—both
financial and, more importantly, ecological—is not a deliberate,
essential part of the deal. But it is a concomitant, and dangerous,
dimension of it. These dangerous consequences are not deliberate
nor are they the result of unconscious behavior. Or, to make the
point clear, there is an apparent disconnect between the plurality
of items of human behavior—deliberate, intentional, unconscious,

whatever—and their aggregated effects. So are there any problematic habits to be found, or just tragically hard-to-avoid consequences of otherwise perfectly sound practices?

I understand by habit repeated, typically unconscious, common practices, in principle amenable to reflective modification. These include cognitive dispositions. Given this model, we can begin to pick out some of the deeper dispositional orientations underlying the toxic consequences we have listed. Greenhouse gas emission is largely the result of burning fossil fuels in our current industrial practices and transportation, as well as our current agricultural practices. Memories of the campfire die hard. The smoke just drifts away. And "away" seems to be somewhere else. There are a couple of ways we can think of what is going wrong here. We could speak of our proclivity for linear thinking, which blinds us to causal loops and cycles. Or we could talk about our tendency to "externalize costs," to use an economic metaphor. Neither of these shapes of thought are bad in and of themselves. Our agricultural practices include raising livestock for food that accounts for about 10 percent of CO_2 emissions worldwide and much higher percentages of methane and nitrous oxide—far worse culprits than CO_2. Cattle farming contributes more to global warming than driving cars. Habits here? We would have to point to deep-rooted (but actually quite recent) assumptions about mobility, about privacy (cars versus trains), and about a carnivorous diet. The growth economy, targeted by people such as Herman Daly, is not hard to deconstruct.[7] It seems to rest on an assumption about the infinite extrapolation of existing trends, as if there were no intrinsic limits to growth. This is the same inductive fallacy that, as Bertrand Russell explained, was dear to the chickens who counted on corn from the farmer in perpetuity. Until the day he wrung their necks. Our commitment to the growth economy is not unconnected to our enjoyment of a consumerist lifestyle and the alignment of our malleable desires with available commodities. Shopping seems so obvious—could it really be a habit?

Finally, what can we say about population growth? Where is the habit here? It would be misleading to describe human reproduction itself as a habit. But the shape and scale of it surely is. Economic development on the whole points in a promising way to lower birth rates, as women have more education and more control over reproduction. And as other social institutions make large families less necessary for security in old age. The problem of population growth does not have to do with a Malthusian explosion, as indeed lemmings do seem to experience, but with a steady rise of global population coupled with a massive increase in the average carbon footprint, as people in China, India, and Brazil understandably aspire to Western lifestyles.

These are just some preliminary forays into some of the affective and cognitive dispositions, and other habits, that lie behind the major causal agents in our climate crisis.

As a philosopher, it would be otiose to try to begin to propose remedies, ways in which habits could be transformed. We all have suggestions, and I am no exception. Instead, I want to take the next step in what I am calling a critique of habit and look at what we might call the evolutionary architecture of our dispositions. Again, I must apologize—I am no expert on evolutionary biology, but even without that expertise observations and questions are surely in order.

I take my orientation from Nietzsche's astonishment that we might believe man's current condition to be the final stage of evolution. This would be another example of our myopia. And, as is well known, he offers a very particular account of the development of our constitutive infection with life-denying ascetic morality and ressentiment as a way of imagining another path toward affirmation and the Übermensch, all the while acknowledging the positive benefits that asceticism brought (including philosophy). I am not proposing to reprise this story, or to engage much with it, but I do want to take seriously something of his genealogical approach, one subsequently taken forward by Foucault.

Current questions about the sustainability of human life on the planet are of two orders: whether in the long term we are not on a path to creating conditions in which we are unlikely to survive. I do not know how likely this is, but it is surely possible. The other order of question has to do with the quality of character of such life. We can imagine shining technocities of Facebook people who have somehow solved the energy problem, the happiness problem (through direct brain stimulation), and so on. We can imagine small pockets of neo-tribal remnants struggling to survive in the face of a general apocalypse. We can imagine a somewhat smaller version of what we have now, run by ruthless military dictatorships. And so on. But we are deluded if we imagine a global expansion of the Western lifestyle. All this is very different from the leisure society some imagined in the '60s, in which technology would free us from brutal labor. What is so distinctive is the eco-biological level at which so much of what is being discussed is taking place. So much of what drives us forward are basic instincts (survival, food, shelter, pleasure, security, recognition) mediated by the social and historical accretions of culture and habit. To the extent that this aggregation of dispositions is the problem, or an important part of the problem, we need to be thinking about how we acquired some of the apparatus we take for granted. Without wanting to endorse Freud's metaphysics, I am convinced by his claim that our current human constitution needs to be understood historically in terms of the pressures faced by our predecessors.

There are no guarantees of harmonious development between biological and social evolution. It is a commonplace that our technological development has outstripped our institutional capacity to regulate it. AK47s are widely available to tribal boy soldiers with little defense against the will of their superiors. Male aggression may be innate, but the consequences of a gun battle in a downtown bar are very different from those of a fistfight. And, of course, anthropologists would contest the premise of innate aggression (see, e.g., the Bishnoi of Northern India). No doubt there have been and still are circumstances and perhaps whole societies in which the capacity to respond quickly with lethal aggression is useful. But that is an argument for minimizing those circumstances and changing the societies in which such habits make sense. A critique of habit would develop this thought.

Time and circumstance teach strange lessons. We can plausibly assume that many of the habits we guard fiercely were developed as legitimate responses to conditions that no longer obtain. Large families do make sense when there is high infant mortality and insecurity in one's old age. Idling at traffic lights (to return to an earlier example) did make sense at a time when cars started less reliably, and it did not matter so much when there were fewer cars and less pressure on the environment. Engines have changed, and so has the situation.

Externalization is surely an aspect of a vitally important capacity—that of focusing on the task at hand, getting a job done. If we always had to worry about the consequences of what we do, many of which are unpredictable, our agency would be compromised. We first peel the orange and then worry about what to do with the peel. So many obvious everyday examples make it clear that externalization is in and of itself a perfectly good habit. It is scale and context that makes it toxic. But it is all too easy not to adjust for scale. It is perhaps worth noting just how successfully right-wing political rhetoric exploits just these gaps of scope, scale, and context. In the United States, I am thinking of the gun lobby, hostility toward the Environmental Protection Agency, and to taxation, and of deregulation advocates generally. Atavistic schemas of personal freedom eventually get wheeled out to justify protecting the free speech of corporations under the First Amendment. My point is that these schemas, often shapes of embodiment (wholeness, integrity, independence, purity), which most likely guide and ground many a disposition, are not bad in themselves but are vulnerable to being appropriated by forces hostile to collective solutions to common problems.

It is worth mentioning here the work of George Lakoff on framing, where he deals with the political power that comes from effective framing of issues and debates.[8] He claims particularly that

Republicans frame issues in terms of the strong father, while Democrats appeal to the nurturing parent. Too often the former schema trumps the latter. If he is right, these schemas are unconscious patterns of thought, cognitive habit structures if you like, association with which often wins consent. If environmental regulation, for example, is successfully associated with castration, who would want it?

I am against biological and sociobiological reductionism as much as the next person. We cannot take for granted what is biological, what is cultural, and so on. But the current state of our terrestrial existence is such that an investigation of our biological history and its potential plasticity is increasingly urgent. It is no accident that it was the Deleuze of *Anti-Oedipus* who said that we do not know what our bodies are capable of.[9] His position lines up with that of Lakoff in wondering whether we might not need to escape from the dominance of the father figure, both law and protector, whose hegemony may now be outdated.

What is interesting about these cases is that they relate not strictly to biology but to vital social relations between family members. It is no accident that Freud understood religion as a displacement of infantile dependence on the father. And again, there are many who reject climate disaster in the belief that they are somehow protected through faith.[10]

There are many other cognitive dispositions that, like Newton's laws, may serve us well enough under restricted conditions but which turned loose on a macro scale serve us ill. Our ability to assess risk, for example, is notoriously flawed. We allow ourselves to discount risk in situations where the risk itself is uncertain or the time long distant. (Scientists disagree about global warming—let us wait and see.) And we often discount catastrophic dangers of low probability. Each of these has a direct impact on our thinking about anthropogenic climate change, as do those we listed earlier—projective thinking, magical thinking (Jesus will return to save us), and so on. Climate scientists have proposed, as an antidote to our seeming willingness to discount risk, the precautionary principle, through which we would take a sane attitude to uncertain risks that action now would avoid. Our tendency toward mimetic sociality, as René Girard would have it, imitating our neighbors, wanting the sort of things they want, may well serve social solidarity. But when fed into a rabidly consumerist society, it is a recipe for a wholly artificial and unsustainable assault on natural capital.

The thrust of what I am gesturing toward here is this: we have inherited a range of cognitive, affective, practical, and social dispositions, both as humans and as members of this or that society, many of which we think of as natural and impervious to change, and many of which had or have a perfectly sound basis but whose scope of application has been lost touch with if ever we knew it. These have especially to do with security, safety, and social existence. And once detached from their scope of reasonable application, they can turn toxic. A critique of habit, in this sense, would take seriously trying to unpack what is truly biological, and what is perhaps open to being reinscribed in different habits, or habits constrained by a different sense of their proper scope. Part of the point of this exercise would be to get on the table some of our sacred cows. I have some very green friends for whom unlimited air travel is just part and parcel of keeping connected with family members spread across the continent. But part of the reason they are so spread is cheap air travel. And as a global academic, I share this "mobile imaginary." I half-heartedly embrace Skype, and yet it is not hard to see that this will one day become a luxury again for elites. Or will there be simu-travel, in which we manipulate fully responsive, distant body replicas, avoiding airport delays and so on?

The Productivity of Habit

If these comments have taken a hard look at the shortcomings of habits, we must also remark on what we might call their productivity. We said earlier that without habit, we would not survive. But that is far too thin a justification of habit. Imagine you are listening to a piano virtuoso playing Bach. You marvel at his touch, his timing, his passion, his sensitivity, and so on. All these are made possible by the fact that he can play the piano, that he has practiced for years, that his fingers automatically go where they are supposed to go, that he knows this piece "by heart," and so on. Or think of the writer, who first learns those "habits" we call individual words (lies, Nietzsche calls them, anthropomorphisms), then grammar, then a certain control, then, perhaps, a voice. At every level a new kind of creativity arises, habits build on and are nested in earlier habits. The same can be said more broadly for complex social and economic structures. Predictable patterns of behavior on the part of others make it possible to plan and carry out more complex acts that presuppose them. Often, these are habitual. If I go out for a meal, I get dressed with movements long practiced, I negotiate the stairs, I lock the door, I walk down the street, I find the restaurant without thinking, I sit down, I order my usual dish, and so on—a nested sequence of habits that makes the outcome possible. If I decide to cook at home and invent a new dish, I will bring all my culinary skills to bear on the new idea. Creativity and complexity presuppose many layers of habit, even as they break with it. And even though there may be more efficient ways of doing some of the things we do, it is the interconnectedness of our behavior that often sustains even downright foolish habits. Better the devil you know, as they say.

Ontological Habit

My "critique of habit" is a placeholder for something more serious. I would want to work with Bourdieu and his sense of habitus, as well as Merleau-Ponty, and Foucault to craft a sense of embodiment as a site of habit sedimentation, through adaptation, trauma, and struggle.[11] And if there is something called critical evolutionary biology, I need to read it. I have perhaps played fast and loose with the very idea of habit, happy to provide a definition but never quite sure whether I am forcing language to accommodate examples. The point of such a critique would be to try to locate the points at which what we may take to be essential to being human, or essential to how we view the shape of our happiness, has been formed, and may in principle be reformed. It would further locate the kinds of investment we may have in our current economies of embodiment, relation, desire, and so forth. (This "we" is, of course, wholly contestable!)

Meanwhile, I return to what will for some be an implausible possibility, namely, that we might revisit Heidegger's later writings for some guidance on how to reinhabit the earth. I say "implausible" for reasons already mentioned. Heidegger keeps his distance from any straightforwardly material understanding of terms such as "earth" or "dwelling." The evil and destruction of war, death camps, and nuclear weapons are for him "more of the same," confirmations of a preexisting dispensation of being, variously called *machenschaft*, or technology.[12] What I find nonetheless intriguing is the thought that what he is describing as a different way of understanding being might adumbrate, be translated into, reinforce, or parallel the kind of change we need. I am encouraged by Nietzsche's prescient thought that the different metaphysical options we have are, at base, different "interpretations of the body."[13] Is this the kind of bridge between materiality and ontology that Heidegger could endorse?

What are the key elements of his position? What hasn't changed from *Being and Time* is the thought that we do not understand that the way we interpret Being is (just) one way among others. This blindness, our forgetfulness of Being, is long-standing. And he finds traces of an awareness of it as far back as Plato. Arguably, the reason for our lack of awareness is that we cannot just throw a switch and change the settings. We are dealing with the workings of history. And a rejection both of Nietzsche's highlighting of will and the politics of the Third Reich is what shapes Heidegger's Nietzsche lectures in the '30s. A key claim of *Contributions* is that we can at best prepare the way for a possible transformation, for the return of the gods who, as Hölderlin put it, have fled. We find ourselves caught up in an englobing system and vision, technological enframing, which hardly allows other voices to be heard. When we truly understand the level at which this is operating, we realize that technology, far from itself being something "technological," has a deep, one is tempted to say, *spiritual* significance, even in its negative re-alization. And this very realization allows us to imagine something different. It is at this point that Heidegger refers to Hölderlin's claim that "where *danger* is, grows the saving *power* also."[14] Heidegger imagines a new "free relation" to technology, by which I believe he means understanding it for what it is, allowing for noninstrumental relations, and being able to act instrumentally without subscribing to its world vision.

All these claims so far can be allowed to spawn analogs of a sort for our thinking about global warming, climate change, and so on. There is clearly no act of will that will fix things, not just because we have no führer with that power but because there is no single act or determinate set of acts to perform. There is a temporal urgency that Heidegger does not accommodate. He may agree that "time is of the essence" but in a rather different sense. And we would want to think of "preparing the way" not simply in terms of offering sem-inars but inventing, investing, trying out new possibilities, model-ing quite dramatic changes, and so forth. But the writing and talking is vital, because language is a critical crossroads for working through where we are at and articulating where we want to go, both in the sense of how we understand our relation to language and also which concepts and words we privilege. And I believe that it is here that we can find the point of Möbian breakthrough that I promised, the point that the two sides of the Möbius strip—let us say, materiality and spirituality—bleed through to each other, short-circuiting the long road of connectedness.

At so many points, and in so many ways, Heidegger tries to wean us off the idea of being in control, the idea that we are autonomous nonrelational agents that can take up the tools in front of us and deploy them. Sometimes, of course, this is just what happens; *Be-ing and Time* highlights our relation to tools by way of contrast with things just present-at-hand. But when it comes to language, art, other humans, our relation to Being, this model is hopelessly misleading. In many of his essays on language, for example, the mes-sage Heidegger drives home is that language cannot just be thought of instrumentally (nor as mere communication, nor expression) because all of these accounts come onto the scene too late. Human beings are, if you like, co-constituted with and by language. Acting and thinking in a way that refuses to see this is indeed an option but a veritable blind alley. Naive agency understands my relation with what I deal with as an external relation when it is, in fact, internal or, more carefully, cannot be thought in inside/outside terms. Instead of seeing ourselves as wielding language, we need to listen to the voice of language, to bid words to come and commend them to the world in which they appear. He sees the best poetry as setting things within the fourfold, of earth and sky, mortals and gods. And when Trakl speaks of the tree of graces, it is of a thing that connects earth and sky in an arc, that captures a certain difference, or rift, that marks the continuing dynamic connectedness between thing and world, never wholly stabilized.[15] And so on. What are we to make of this?

One's first reaction is that it reflects a repeated refusal to deal with the gritty sensuous materiality of things, with the plight of the earth. Where do oceanic dead zones fit into this picture, or mass species extinction? But it is worth recalling how Marx understood materialism, as focusing on (actual) relations of production (between humans), relations of forces, differential relations to capital, and so on. It is not that Marx was insensitive to soot and disease and poverty, far from it, but materialism meant focusing in on the effective structures and relations that drove the modern world, as opposed to Hegel's seeming willingness to speak of spirit. In so doing, he stood Hegel on his head!

Can we say something similar about Heidegger? Can we describe the difference between thing and world, or man's relation to language, or the way we understand our sedimented habits as *material* relations? This is what I would like to claim. Our real presence in the world is changed by how we understand our relation to our own social and biological constitution. Understanding that relation after having gone through something like a critique of habit can change the content as well as the character of our agency, just as coming to see the limits of linear thinking can encourage recycling. But this is to make the very general point that even complex, seemingly abstract patterns of thought can be materially efficacious. What of the specific contribution of Heidegger's thought to a renewal of the shape of our dwelling on the planet, reinhabiting the earth?

Given the racist and biological ideologies of the Third Reich, Heidegger was understandably cautious about snuggling up too close to biological science. If anything, biology has the same impact on life, cognitively, as technology does on nature. Heidegger understands nature essentially as *phusis*, as an ongoing, upsurging, overflowing event or force. It can welcome us, threaten us, even overwhelm us. Technology *sets upon* nature, challenges it forth, as he puts it, and attempts to turn it into a standing reserve. In a windmill, we are still dependent on the vagaries of the wind. With the national power grid, we declare victory. Nature can be turned on and off. Poetry, not to mention hiking in the hills, reminds us of a different, less willful relation.

And just as Heidegger resists the technological transformation of our whole way of thinking about nature, he will also resist what we might call the anthropologizing of the animal that would come from just allowing that they each, too, have their own world. Instead, he claims that the very idea of world needs to be used cautiously. Animals are worldless, or poor in world. We need to be cautious about subscribing to Rilke's sense of how the world is open to the animal. Ideas of truth and freedom do not simply transfer from man to animal.

His introduction of the fourfold, the *Gevierte*, can perhaps best be understood as a way of providing an alternative "frame of reference" to technology's *Gestell* (enframing), one that displaces man from the center while indicating the key dimensions of "significance" within which we can understand the things we encounter. The implication is that this background needs to be supplied for things to resist a technological reduction.[16]

Is the language in which this is couched at all helpful for reinhabiting the earth? What concretely can we take from "poetically man dwells"? It is pretty clear that Heidegger is diagnosing, in the shape of *Gestell*, what I would call an ontological habit, a socially, historically, and existentially sedimented way of understanding and relating to the world, with particular impact on the natural world to which we have independent, though fragile, access. The change in the Rhine when we come to think of it in terms of hydroelectric power—rather than, more poetically, as the heart of a nation—is real and profound. Heidegger also explains how this habit is hard to change. As with any dispensation of being, it lies, largely invisibly, in the background. It involves our attitudes, our behavior, our cognitive and affective dispositions, and even the language we use to talk about the world and the things in it. These are all habits, albeit nested, interconnected, and in principle capable of change, even if the obstacles to such change are huge. And the change in question is one in which there would be a certain shift in disclosedness, in truth.

So is there anything missing from Heidegger's account? I have tried to argue that (following Marx) if we understand effective relations between things as such as material (for Marx, relations of production), then Heidegger is talking materially, not just spiritually or ontologically. In truth, I suspect this language breaks down. But I am also tempted by the thought that we can perhaps do to Heidegger what Marx did to Hegel and interpret his claims in a more straightforwardly material way. I am not sure if I want to do this. But there is something missing in Heidegger's account that needs to be remedied. We are said to be witnessing the Sixth Great Extinction. We are creating conditions in which climate change will outpace the capacity of much terrestrial life to adapt through normal evolutionary mechanisms. We are arguably heading for a world in which analogs of biblical plagues will fall upon us—mass forced migration, famine, resource wars, political regression, and so on, as well as drug-resistant bacteria and other obstacles. If we can connect the dots here, these negative phenomena are the causal consequences of our ontological habits. The desire for control has swung out of control. We have not lost the capacity to listen, to look at the big picture, to reimagine the world. But these voices are not running the show.

Heidegger seems to be telling us that the "danger" from which we need to be saved is spiritual, that we are losing touch with the truth of what it is to be alive. Even if, as Heidegger tells us, science does not "think," it surely sets us thinking when it tragically plots our fragile future. But what Heidegger makes me ask is whether we are not now confronted with a strange convergence between a spiritual and a material collapse, such that the capacity of the planet to sustain life as we know it, and our extraordinary hubris as a species in treating it as *our* oyster, might be on a real historical collision course. Our deep habits might find us lined up in an underground cave watching shadows on the wall not merely in an allegorical sense, sheltering from the truth, but taking shelter from the storm.

What I am proposing, then, is that Heidegger offers us at least an example of how to excavate and articulate the level of habit that may need to be reworked in any future reinhabiting of the earth. And I alluded to our historically sedimented bodily dispositions and to the anthropogenic toxification of the planet to suggest that we may need to set aside the opposition between the material and the spiritual. The Möbius strip showed us the two could be thought on the same plane. It may be that our time can be likened to a pin being stuck through from one side to the other.

Itching and Scratching

But aren't these dire predictions about the end of the world just alarmist exaggerations? A pragmatist friend of mine once advised me: Don't scratch where it doesn't itch. His comment really got under my skin, and I have been scratching at it ever since. The problem, of course, is that the lack of itching may be the problem, like the proverbial dog that did not bark in the night. I am grateful that when summer ticks bury their jaws in my leg, they instigate an itch.

In truth, philosophy has often dealt with problems like this. Injustice is a major issue, but not recognizing something as injustice is worse. Oppression is disguised in many ways. Itches can be masked by the anaesthetics of false consciousness, or complacency, or blindness. The problem with Plato's cave dwellers was not that they were disturbed by the shadows on which they feasted their eyes but that they didn't even know that they were shadows. And, famously, Heidegger begins *Being and Time* tasked with resuscitating the question of the meaning of being—which no longer even strikes us as a question.[17]

The proper response to my skeptical friend must surely be that philosophy itself is threatened if we scratch only where it itches. There are, of course, plenty of practical challenges that urgently need addressing: poverty, war, oppression, preventable disease, and so on. But even then the real problem may well lie in our being too comfortable with the conditions that sustain them: our comfortable lives may be itch-free, even as we feed the bonfires of catastrophe with the by-products of our normal practices. Under such circumstances, calls for revolution understandably fall on deaf ears. And those who have the itch to do something are stymied by the readily available models of revolution. Cutting off the king's head and marching down the street with it stuck on a pike, however theatrically satisfying, is no longer an effective path to the change we need, though one should not doubt the power of televised terrorist beheadings, even today.

The obvious response to this way of framing the problems that face the planet—notably the prospect of catastrophic climate change—is that the problem is not that we aren't itching but that we *are* itching and scratching. But scratching only brings temporary superficial relief, and as with poison ivy, scratching can make it worse. It has been said that philosophy boils no cabbages. Yet it is an activity, not just a collection of texts. It involves arguments, interventions, critiques, imagining alternatives, inventing new concepts, and revising or reviving old ones. Philosophy is called for by specific situations and circumstances, and it may need first to crystallize conflicts in order to resolve them. But it is not directly practical. When Marx said that "[t]he philosophers have only interpreted the world in various ways; the point is to change it," this is not to condemn philosophy but to say that it is not enough.[18] Reinterpreting the world may be a necessary condition for change, even if it is not sufficient; that is the premise of this book. But even "interpretation" needs interpretation. It is not enough that we devise new ideas, or stories, or models and set them before us on the shelf. We need to devise ways in which they open up new forms of dwelling, ways of reinhabiting the earth. To speak of "dwelling" is a placeholder in a strange sense. It attempts to capture the shapes of being in the world, our "forms of life." These shapes are concrete and practical—housing, travel, food, patterns of exchange and production—and they are each sites where ideas are exercised, explored, and embodied.

Does philosophy, like the owl of Minerva, fly only at dusk? Can it only reflect on what has already happened, sorting out the accounts? If there is more in heaven and earth than is dreamt of in our philosophy, does not that consign philosophy to an ancillary role, picking up the pieces after the wagon train of history has passed? It may be that time has always been out of joint, and yet this time may be especially so. Dusk today does not herald a predictable new dawn. The story to be gleaned from the rearview mirror is a dog's breakfast chiaroscuro. Glorious human achievements mixed with folly and horror. And we have long since lost faith in the grand narratives that could justify man's ways to man. One man's progress is another man's genocide. One could at least imagine some sort of reconciliation with our violent past if there were the prospect of a rosy dawn. Words like "sacrifice," distasteful as they are, might find some purchase. Instead, however, we find ourselves inheriting a past that will likely usher in a catastrophic future. The solace of sacrifice in the name of progress has evaporated.

PART

I

Econvergences

CHAPTER

1

On the Way to Econstruction

The future can only be anticipated in the form of an absolute danger.
—*Derrida*, Of Grammatology

Nostalgia runs all through this society—fortunately, for it may be
our only hope of salvation.
—*Donald Worster*, The Wealth of Nature

I have long engaged in productive dialogue—first with phenomenology, and then deconstruction—while keeping my environmental concerns on the backburner. While I have no illusions about our capacity to entertain quite incompatible lines of thought, the question I want to pose here is whether green deconstruction need be an oxymoron. Or whether, conversely, a living and developing deconstruction might find itself quite at home thinking through the quandaries of environmental concern. Is econstruction simply a monster? Or might not environmentalism provoke a healthy materialistic mutation within deconstruction?[1]

Let me begin by amplifying the two questions that will help move us forward.

First guiding question: *Might a living and developing deconstruction find itself quite at home thinking through the quandaries of environmental concern?*

29

We may think of deconstruction as a kind of antinaturalism, one that takes seriously complex nonnatural structures and relations—from logical aporiae to undecidability and infinite responsibility. We may think that these are a far cry from the accelerating rate of species loss, global warming, deforestation, the depletion of natural resources, the hole in the ozone layer, the sucking dry of deep aquifers, the dangers of catastrophic changes in weather patterns, and so on. These all seem like very real concrete phenomena, dangers we do not need philosophy, let alone econstruction, to know we need to address. And each one of these "dangers" is not simply a sharply defined object of knowledge but an issue, a concern, a site of complex relationality. As much as these are what we call "natural" phenomena—who could doubt the naturalness of a tornado that uproots trees, sucks windows out of skyscrapers, throws cars across town—what is at issue for us is to understand their cause, the probability of them occurring, the true danger they do pose, how far we are really responsible, and how they might be averted. In other words, while it is a crucial matter of intellectual integrity to acknowledge that we are talking about real phenomena whose reality can be measured in the hard currency of death and destruction, these phenomena are *issues* for us because of further questions about the adequacy of our knowledge and control, the models we deploy to engage with these phenomena, the social practices that enable and disable appropriate responses to these problems, and the deep difficulties we face in trying to think through the contradictions they present. The irrepressible reality of these phenomena does not in the end tell us what to do, or how to think about them. To the extent that deconstruction trades in this complexity, it might be thought to be just what environmentalism needed.[2]

Second guiding question: *Might not environmentalism provoke a certain materialistic mutation within deconstruction?*

A new materialism? Where it has not been a formal charge, it has often remained a suspicion—that deconstruction reflects a kind of academic detachment from the real. Reading Derrida's "there is nothing outside the text" literally reinforces this suspicion. We might suppose that this would present an obstacle to the fully blown materialism to which any environmentalism is surely committed.

Long ago, in the afterword to *Limited Inc.* ("An Ethics of Discussion")[3] and in many other places, Derrida was at pains to insist that "text" does not mean words on a page, and that "con-text" might have been less misleading. And he went on to add that while all meaning is contextual, no context is ever fully saturated or complete. Central to deconstruction, too, is a refusal of any absolute distinction between meaning and force, expression and indication, intentional and causal. What we think of as a privileged realm of human meaning-giving activity is always exposed, internally and externally, to what it seeks to exclude—which we could call the real, the outside, even nature. We *could* call that exposure the materiality of meaning. This word "material" is worth pausing over, because it is the source of much misunderstanding. On one reading, to insist on the material is a kind of reductionism, one that begins with complex meaningful, even spiritual, phenomena and reduces them to cruder material forces. "Freud reduces everything to sex" would be an example. But a quite different reading would insist on the multiplicity, complexity, and multidimensionality of material forces. This approach, for example, would agree with Marx when he understood "relations of production" as material forces, even if it could not subscribe to a single root-cause foundationalism, a "last analysis." Material forces here would cover a wide range of phenomena, from the effect of automotive lead poisoning on child brain development, to the role of economic interest in a discretionary war. And the point of going *material* is not reductive at all, not an attempt to reduce apparent complexity to some single underlying material cause but rather to disclose and explore the wealth of the real.

Deconstruction had its roots in Derrida's taking up of an early materialist critical reassessment of Husserl by Tran Duc Thao. It is true that deconstruction focuses less on the *matter* of materiality than on the significance of the irreducibility of the material, the limits and paradoxes of ideality. We might say that deconstruction is a strategic rather than a substantive materialism. It acknowledges and welcomes the interruption wrought by an excluded materiality. In this way, we may suppose, it is well positioned to help us think environmentalism not as a positive science (as Husserl called phenomenology) but as a challenge to any science blind to its necessary ideality.

The remarks we quoted earlier ("There is nothing outside the text") lead Derrida to a discussion of the logic of supplementarity, in which something supposedly complete (e.g., nature) needs a supplement (writing) that will expose the original completeness as a lack.[4] This reminds us, if we needed reminding, that environmentalism cannot just be the champion of a lost purity of Nature. It has to reckon with the inseparability of our thinking about Nature from the theme of loss, alongside the *real* loss reflected in the avalanche of species that are becoming extinct every *day*. In his book *The End of Nature*, for example, Bill McKibben argues that manmade atmospheric changes alone mean that nothing on earth is any longer in its natural state, unaffected by man's presence. But it would be foolish to conclude that because there is no pure nature anymore (if there ever was), we do not need to make distinctions between different kinds or levels of loss. Donald Worster opens his book *The Wealth of Nature* with an unashamedly nostalgic evocation of the natural abundance that was once America. We will consider later how to evaluate this attitude.

Animal Rights and Environmentalism

I am not claiming that Derrida explicitly saw environmentalism as the next step for deconstruction. It could be argued,[5] indeed, that the direction he does take cuts against environmental concerns. In an extension of our broad responsibility for the human other, he has on a number of occasions attempted to articulate a face-to-face relation to his cat.[6] There is, however, a well-documented tension between those who take up questions of an individual animal's rights or their well-being and those who pursue environmental issues.

The animal rights advocate will rescue the bison trapped on the ice, while the environmentalist will think of the bear and her cubs that depend for their survival on such unfortunate accidents. Derrida does indeed problematize the ethical focus on the privileged individual (his cat). But he does so by asking how we can justify ignoring all the other (individual) cats; he does not talk about mice or birds. Or the cats that the Egyptians domesticated to keep snakes out of the house. In taking this path, Derrida follows Levinas in seeing the movement from the ethical to the political in terms of the importance of the *third*, who is always implicit in the otherwise privileged face-to-face relation, muddying the waters. Still intact is the implication of a potential personal relationship with a discrete individual—that the other typically has a face, and that it is hard to know how to deal with the *many* others whose faces are indistinguishable from those we happen to meet. Arguably, there is a residual humanism in this approach, as with Levinas. The first stage of otherness, at least, opens us up to creatures a bit like us, or those with whom we share our lives. Moreover, the ethical focus on the asymmetry of obligation, on the gift that seeks no return, one that is even compromised by that possibility, suggests an awkwardness in thinking the deep interdependency with other life-forms that is surely our condition. But we may properly ask whether Derrida does not offer us elsewhere the resources to take our thinking further.

This question of humanism arguably taints even the very word "environment" in its suggestion that we concern ourselves with what surrounds *us*.[7] Strategically, and historically, there is some justification for this focus. In many ways, it is our blindness to and lack of interest in the impact of our ways on "what surrounds us" that is the source of the problem. The problem is no longer that we are polluting "our surroundings" but that we are transforming the earth in ways that are deleterious for it, and not just for us. Derrida did not write much, if at all, about the environment. In *Specters of Marx*, he writes of both *earth* and *world*. But in each case, the primary focus is on human beings. Thus, "[N]ever have violence, inequality . . . affected so many human beings in the history of the earth and of humanity . . . never have so many men, women and children been subjugated, starved or exterminated on the earth."[8]

It is at this point that he pauses to set aside "the indissociable question of . . . so-called 'animal' life" in a reprise of Levinas's move that treats the face of the animal as derivative from that of the human, casting serious doubt on issuing any credentials to the snake. The chapter in which this passage occurs (Chapter 3, "Wears and Tears: Tableau of an Ageless World") begins: "*The time is out of joint. The world is going badly.*" Here the "world" is that of the New World Order proclaimed by liberal triumphalism. "World" here means the human project (of freedom and enlightenment). Derrida need not be blamed for framing things in this way. He is precisely responding to an exceedingly human way of understanding planetary history. And it is in this context that he will enumerate the ten plagues of this "new world order."[9] These plagues both threaten the attainment of such an order and also threaten the credibility of the very idea of such an order. While the seven biblical plagues of the book of Revelation[10] are eerily close to some of the ecological threats that face us today, Derrida's plagues are all plagues that beset human institutions, especially those that threaten the possibility of realizing such fundamental values as peace and justice. However, we do not need to apologize for suggesting that the destruction of the earth should be listed as the eleventh plague.[11] Not only does it have an importance at least as great as the other ten, but its interruption of the human institutional space that the others occupy is something of a second-order plague. Derrida speaks of messianicity without messianism as the unpredictable arrival of something or someone, unpredictable both in its timing and in its outcome. The implication is that this arrival will be our salvation, or at least be positive. But if we set aside that presumption for the moment, the destruction of the planet, the destabilization of its sustaining life processes is an intrusion on our human project quite as powerful as the appearance of the refugee on our doorstep. Environmental destruction gives us a wake-up call of epic proportions and is surely a candidate for the status of *arrivant*. For it arrives *as* something that has been excluded, much as Freud describes the return of the repressed. In the "Exergue" of *Of Grammatology*, Derrida writes that the future can only be anticipated as an absolute danger. With that seemingly apocalyptic remark, he was drawing attention to the shift from what we might call a phenomenological model of meaning (which would privilege the voice) to one centered on *writing*, in which meaning (and history, and the future) would escape the control of guiding human intentionality. The emergence of ineliminable paradox, and of structures of aporia, reflects the absence of any overall synthesizing power. The prospect of the collapse of that assurance is akin to what Nietzsche called the death of God, and, from the point of view of the project of meaning, it can only appear to be terrifying, out of control, "absolute danger." Writing infects meaning from an exteriority that turns out to be immanent to meaning and not exterior at all. Something analogous, at a second level, might be thought to be happening with current environmental threats. They seem to be coming from the outside. And yet they reflect the structure of what the CIA calls blowback, the facial smear that follows spitting into the wind. What seems to come from outside are, in fact, the cumulative consequences of our own actions. As ye sow, so shall ye reap. What goes around comes around. Here we meet a parallel with part of Derrida's analysis of 9/11, in which he questions the idea that these attacks came from "without."[12] This sense that the "outside" may reflect a certain blind construction of inside/outside that needs to be reevaluated parallels the sense that the environment may not be, as we suppose, what surrounds us; it may equally be what pervades us or what has a precarious life of its own.

Externalities

In various early essays,[13] Derrida writes of the need for a double strategy: immanent critique and the step beyond, working within the closure of metaphysics and attempting a creative leap outside its borders. And in many other places, he will destabilize the assured boundary between inside and outside, not least in various performative hesitations over the idea of a preface ("Hors Livre"). We might reconstruct what is going on here in the following terms: when we think naturally, literally, we suppose that the inside is inside and the outside is outside. But as soon as we realize that inside/ outside operate within a signifying space, the clarity of that distinction starts to break down and so, too, do the moves (metaphysical, ethical, political) that presuppose the stable operation of this distinction. Interestingly enough, the very distinction between the way inside/outside operates at the level of nature and that of signification itself breaks down. We only need to ask ourselves about how this distinction applies to a living being to find ourselves at sea. Not only is it dynamic (breathing, excretion, digestion, etc.) but it is also multidimensional. Think of the ways in which we appropriate intimate, proximate, and dwelling spaces as part of our "interiority." This makes clear that something like the structure of the trace is already operative "in nature," if we can say that. It does not wait for language to come on the scene.

Equally clearly, the demand for operative demarcations between inside and outside continues unimpeded, tied up, as it is, with questions about responsibility, respect, and integrity but also with property, exchange, and profit. It is by drawing on these latter considerations, economic in a literal sense, that a certain deconstructive lens can shed light on our environmental predicament.

It may be that what we are tempted to think of as a metaphysical illusion—the clear demarcation of self and other, inside and outside—is, as Kant suggests elsewhere, not simply a mistake but a misunderstanding about the *scope* of a distinction that, up to a point or under certain conditions, we cannot avoid making. Although one might think that our entering into relations of exchange both with others and with the world would compromise any strong sense of inside/outside, it could be argued that it is precisely by such a distinction that such exchanges are regulated, subject to some sort of law. They are thus *enabled* in an economy in which time, effort, and energy are productively directed and calculated. It is such an economic space that, as I understand it, Heidegger describes it in

Being and Time as the space of the "in order to," a causal/practical nexus.[14] Because all living beings, including humans, are engaged in other modes and forms of exchange, as well as in relations hard to reduce to exchange, the inside/outside relation turns out to be multiply laminated and undecidable.

This helps illuminate a fundamental driving force behind our environmental crisis. For what is true of personal exchange relations between humans is carried over into those legal persons we call businesses. Whether we are dealing with extractive industries, where the primary "exchange" is with nature, or with processing or service industries, the basic law of success is to maximize profit and minimize cost. And the secret of minimizing costs is to externalize them as much as possible.[15] "Externalize" here means passing on the costs to someone else or, even better, to *something* that will not notice this happening. This is the mechanism that links profit to pollution, pumping toxic waste into rivers, lead into the atmosphere, and garbage into landfills. It is not at all mysterious. Many people who change their own motor oil pour the old oil down storm drains, where it enters the waste water system with dire consequences. This externalizes the cost of the transaction that would otherwise involve a trip to the recycling depot. The natural world, especially air and water, is the prime candidate for being the recipient of these externalities. Scaled up, shapes of practice that were once harmless turn dangerous, even lethal. When no one lives downstream, pissing into the river has a negligible effect. But strings of townships engaged in such practices on a macro scale on the same river will destroy the life in it. What has happened? What has happened is that there is no downstream anymore, no outside, no elsewhere. And yet individuals and businesses continue to survive and flourish to the extent that they continue to maximize the externalization of their costs. Without tough and enforced environmental regulations, it is cheaper to pollute and risk fines.

On the model I am sketching here, the business/technological/industrial world is devoted to the creation of what I would call *toxic identities*, which flourish to the extent that they can excrete their waste products into a relatively cost-free outside. Another word for that outside is the "environment." We may think that ethical and political implications flow from this account, and they do. But the analysis of profit-making entities as cost-externalizing devices is meant to be metaphysically neutral! It may be the case that for this or that business, recycling will pay off—for example, recovering precious metals from one's waste. Whether it does is a matter of chance. The iron law of competition requires that costs be maximally externalized. Businesses that do not do this will go to the wall.

On this analysis, we can see how the inside/outside distinction is *not* a natural property, but a highly constructed one, essential to the persistence of those artificial entities we call businesses, and even to the exchange relations we enter into as individuals. Inside/outside is in this sense, as Derrida would say, undecidable.[16] But the effort to decide it, to determine it in this or that case, is central both to the current ways in which we conduct our economic life and also to the environmental destruction that comes in its wake.

Unsettling the Present

It soon becomes apparent that this reconfiguration of outside as inside is quite as much a temporal phenomena as a spatial one. Here Derrida's sense of the past haunting us, especially that past that we thought we could bury (e.g., the thinking of Marx), is apposite. We frequently hear the claim that if we stopped all carbon emissions today, the atmosphere would continue to warm into the foreseeable future from the effects of past CO_2 emissions. In environmental thinking, there is no shortage of concrete ways of understanding the idea that our time is out of joint. It is a truism that the present is overlaid by past and future. This is enough to wean us off an assurance that we can seal off this or some other present as a secure basis of meaning. But not only, as we have suggested, are accounts with the past far from settled; we are equally unable to accurately predict vital aspects of the future. And the various scenarios that are being projected set the stage for wars, some ideological, others real.[17]

Never has the future been so indeterminate, and this epistemological shortfall has a dramatic consequence. There are many destructive environmental trends about which we will not be able to be *completely certain* until it is too late to do anything. It is remotely possible that current global warming, for example, is just part of a natural cycle. Those who fear otherwise argue, on the basis of the precautionary principle, that we need to act as if it were true, for to wait to find out could be fatal—a bit like Pascal's wager in reverse. It would be foolish, for example, to set the bar of proof so high that one would not stop smoking until one had personal proof of its carcinogenic potential—perhaps a small tumor.

The Reservoir of Difference

One of the most shocking environmental statistics is the rate of species loss. This rate is estimated—and there can only be estimates—at between 50 and 150 a day,[18] and it is accelerating. In the course of natural evolutionary change, there will always be creatures and species evolving and dying out. That is not the shock. What is troubling is that the current rate is said to be between one thousand and ten thousand times faster than what would be occurring in the absence of humans. One would need to go back sixty-five million years to when the dinosaurs disappeared to find pale comparisons. By the next century, half of our planet's species will be gone. We are now witnessing the fastest rate of extinction in the history of the earth. Why is this happening? Habitat destruction, overgrazing, logging, water pollution, atmospheric changes, and our general blindness to the consequences of human activity are chiefly responsible. But is there really anything wrong with this? If we visit the paint store and bring back a large number of swatches, don't we expect to discard most of them when we figure out which we need? Could we not think of the superabundance of species in the same way? Aren't many of them just redundant? Evolution began as a biological process, but with humanity it has itself evolved into a new project, the project of freedom. Can there really be any comparison between the value of one creative, self-aware human being and all the beetles on the forest floor?

Luc Ferry's charge of ecofascism[19] against the *soixante-huitards* (the Paris '68ers) is perhaps driven by the sense that the enlightenment project of freedom is under threat. But arguably the threat does not come from a group of French philosophers but from the history that has been made in its name. The value of freedom has become inseparable from the logic of sacrifice by which those who claim to be promoting this value can justify bringing death and destruction to soldiers and innocent civilians alike.[20] Equally, the project of freedom can clearly be used to justify the mass destruction of other living species. If a new order of value begins with the human, then the subhuman becomes at best a resource for the realization of the human project. We might perhaps err on the side of caution in allowing other species to die out, just in case they harbored useful compounds from which we might make lifesaving drugs (Amazonian plants), or were useful in keeping other noxious creatures in check (a few wolves), or brightened up our parks (birds with pleasant songs), or delighted our sense of the diversity of life. But in principle, we should not mourn the loss of the surplus of species that have aimlessly adapted themselves to niches that are disappearing. On this view, we need to stop being sentimental. Things change! Life is flux! It goes without saying that this naive reductive Darwinism, especially when grafted onto the capitalist enterprise, is arguably the most dangerous essentialization in history.

Dependence and Interdependence

What has been called antihumanism is not antihuman. It is rather an attempt to interrogate the privilege of the human, bound as we are to material flows, as the unthematized point of departure for reflection. It is not to deny that (as far as we know) it is only humans that *can* reflect—at least at this level. It is rather to affirm that it may be one of our powers to step back from a narrow understanding of our own privileged position. Too great a self-assurance about our uniqueness, originality, and independence of agency may blind us to the truth of our condition. After both Heidegger and Derrida, we could describe this condition as one of constructive dependency. In each case, a certain decentering of origin creates a space in which a fundamental relational dependency becomes visible. For Heidegger, this first appears in his account of Dasein as being-in-the-world (not as consciousness set against the world) and later as a being whose agency arises through a capacity to respond to what language opens up ("Man speaks in that he responds to language"[21]). For Derrida, "a writer writes in a language and in a logic whose proper system, laws, and life his discourse by definition cannot dominate absolutely."[22] If we were to translate this model into the natural setting, we would have to say that our agency (and identity, and even dominion) rests on a capacity to relate to, and respond to, and negotiate productively with the natural world around us. But what kind of claim is this? Constitutive dependency can be thought of in rudimentary material terms—air, nutrition, sensory stimulation—without which we would not make it through to the next day.

This may sound like an enlightened anthropocentrism, but it is just as much a recognition of a kind of fundamental coexistence with the rest of life. The pleasure we take in the birds and insects and stars is in part a pleasure at their having a certain independence from us! And if we came to recognize other creatures as having an intrinsic value, or a value not dependent simply on their use-value for us, would not this human achievement at least *compete* with the value of that "freedom" by which we think we can license ourselves to sacrifice "subordinate" forms of life?[23] Acknowledgment of the independent significance of essentially interdependent beings with whom we coexist is not a simple thought. It leaves open how we negotiate between species interdependence (we humans *need* bees for pollination) and our care and concern for individual creatures.

As humans, we do not typically set out to kill off entire species (with the exception of some diseases, such as smallpox and malaria, and many species of wolf). Rather, we neglect to protect their conditions of life (habitat). Or we hunt or trap or fish them to extinction.

We do not need to engage in moralism to find this problematic, especially given the scale of extinction to which we have already alluded. But how should we best frame the problem? So far we have proposed two competing frames, the first unashamedly anthropocentric, in which the only limit we might impose on extinction would rest on a species' amenity value to us, plus a bit of a margin for uncertainty, perhaps. The second would see us humans as interdependently coexisting with other species and finding our value as humans to rest on our capacity to celebrate that coexistence and desire to preserve it. I now want to propose something of a basis for this second frame, one that taps into Derrida's articulation of difference, and differentiation, over any essentialist account of identity.

Who Are "We"?

When we look at the stars, we are rightly awed by what we are seeing. When we realize that the light has taken so long to get here that what we are seeing may no longer exist, the astonishment only increases. And when we further reflect that for all their size and magnificence, they cannot see us, and that they most likely are huge clumps of swirling dead matter, we may well turn our astonishment back on ourselves, both as human beings (but also as sentient beings) and simply as living beings. We creatures may be made of stardust, but any living being is arguably of a complexity that inorganic matter cannot approach. If anthropocentrism is a kind of myopia, I would argue that biocentrism is a kind of full-spectrum seeing, a capacity to respond to other life-forms that is only possible for us because we are ourselves life-forms. What other life-forms offer us is the opportunity for the affirmation of difference, other ways of organizing reproductive complexity.

To celebrate the existence of species is to celebrate deep difference. The loss of a species is the loss of an irreplaceable evolutionary history, not to mention a piece of an interconnected ecological puzzle. But it is also the loss of a future possibility of transformation, adaptation, and differentiation. We may be tempted to think of biodiversity as something of decorative significance. Of course, we do and should delight in butterflies and swallows. But much more than that—it is the finite biological pool of the possibilities of life differentiation and transformation. We have no evidence that this occurs anywhere else but here on earth, and it is an adventure of which we are, perhaps obscurely, only a part. John Donne once wrote that "any man's death diminishes me," and the same is true for the deaths of other species. It is not just that their extinction diminishes us. It tears at the web of life to which we each belong.[24]

Obviously part of the point of using this language is to induce a sense of connectedness between human beings and other beings that may not be at the forefront of our everyday experience. Indeed, we try not to share our domestic and intimate lives with noxious bacteria and viruses, with ticks, leeches, headlice, bedbugs, hornets, mosquitos, rodents, spiders, or snakes. Hospitality has its limits. But the argument never was that we might not legitimately want to keep clean and avoid disease; all creatures protect their bodily integrity in that way. The argument is that what we take to be "other"— our categorizing it as utterly alien, as dispensable for our existence, as outside our space of concern—may reflect a myopic prejudice. "We" are human. But equally, "we" are white, "we" are men, we are "semi-affluent English-speaking academics," and so on. Who are we really? "We" mammals? And yet when intent on a frog breathing, we notice the rise and fall of our own chest, but what is it we are registering? Who are "we"? If *you*, my friend, see nothing, are *we* a we? We intellectuals are quite rightly afraid of anthropomorphizing, but perhaps we lean over backward so far that we eliminate a quite legitimate biomorphizing, legitimate because just being alive gives us imaginative and projective access to all kinds of living beings. Even when they are very unlike us in body or habit, we can deploy schematic analogies to forge bridges.

When Derrida speaks of animals, he tends to use scare quotes, not least because the very word (which he plays with as *"animot"*) encourages a homogenizing differentiation from the human, as if being different from us made them similar to one another. Are not humans essentially different from animals? Well, which humans, and which animals? And why do we suppose that there are differences and then *essential* differences?[25] What Derrida is encouraging is a radical de-essentializing and a persistent suspicion of the language of anthropocentric convenience. But we cannot pretend that there is not a play of both analogy and difference in the extension of our ability to respond to each other and to other living beings. This was something of the point of his reading of Levinas in "Violence and Metaphysics."[26] What I propose here is a biocentric pathway for such analogizing, even as we must give our morphological imagination full rein. I do not need to be able to imagine being a six-foot-long squid to respect its right to exist. But when we chart new analogical pathways to other species, do we not need to hold on to a certain stable sense of our own species? Those opposed to Darwin demonstrate a high level of anxiety at this point. And it is worth sharpening what Derrida would call the undecidable dimension of this issue rather than aiming for a premature resolution.

The idea that the human species has a lineage that connects it in evolutionary time with the higher apes is a threat to those with a certain understanding of essentialistic identity, one that cannot allow that something quite new could develop in time from something different. This, however, is a genuine cognitive mistake. We know a butterfly develops from a chrysalis. We know there is a qualitative difference between a bike kit and the fully assembled bike. In neither case is the product compromised by its origin. Rather it is made possible by it. Resistance to evolution is sufficiently explained by the fear that we may not have complete control over those aspects of our animal ancestry that might still lie within us. The need to confirm the boundaries of our own species is also political, in that it serves to unify the various human races under the common legal framework of human rights (which is a good thing). But would this strategic political consideration arise if we accorded apes and monkeys an appropriate respect?[27] Can we imagine a world in which respect for all other beings in their deep singularity was sufficiently firmly ingrained that our own species identity no longer licensed us to kill, mistreat, and neglect others? In his *The Other Heading*, Derrida takes up the question of European privilege, the kind of privilege that would give Europe a preeminence in human history.[28] He concludes that this could only be justified if Europe were to offer unconditional hospitality to the Other, were to welcome the rest of the world. One could rework the form of this argument as a way of reformulating the distinctive privilege of the human—that we alone are perhaps capable *both* of extraordinary destructiveness and blindness to the fate of other beings *and* of the far-reaching responsibility, compassion, and hospitality toward other living beings.[29] Can we seriously claim the privileges of the human when we deal with other creatures in such a *beastly* way?[30]

Justice beyond Representation

In *Force of Law*, Derrida writes: "[T]here is no justice except to the degree that some event is possible which, as event, exceeds calculation, rules, programs, anticipations and so forth. Justice as the experience of absolute alterity is unpresentable."[31] Elsewhere, he speaks of going through the undecidable as the condition for responsibility, that there is no formula, no algorithm that can decide for us. The attempt to mark a space—perhaps the space of the ethical—beyond calculation, beyond representation is a persistent theme of Derrida's later writing. If there are times when one might be forgiven for understanding these remarks in an existential vein, there is nothing quite like the environmental landscape for shaking one from this interpretation.

Going through the undecidable is a multistage process. Even economists disagree enormously about what figures to use when trying to evaluate environmental impacts. I have read reports in which their difficulty in deciding a figure led them to assign a zero value to a certain amenity in their equations, as if zero represented matters more adequately. But after that, environmentalists face a huge dilemma—whether to accept the demand to calculate, to play the economic game, or to hold out for a place at the table for nonquantifiable or

incommensurable values. And, of course, it would be another mistake to "represent" this situation as one in which delicate nature
is being violently trampled on by brutal humanity. There may well
be many conflicting calls on our conscience, each of which resists
quantification.

Undecidability was never offered as a helpful decision procedure,
and we should not be surprised when it opens us up to fractally proliferating spaces. What environmental dilemmas make clear, however, is that these are not the personal dilemmas of an existential
conscience but, dare one say it, the problems posed by the real when
it bites back against the poverty of our attempts adequately to represent it.

The Parliament of the Living as the Democracy-to-Come?

Consider then how one might rework Derrida's idea of a democracy-
to-come within an environmental context.[32] Derrida connects the
ideal of equality within democracy to that of the fraternal bond and
looks to find ways of overcoming that limitation. It is in this context perhaps instructive that in his famous poem, "The Canticle of
Brother Sun," St. Francis of Assisi addresses himself to Brother Sun,
Sister Moon and stars, Brother Wind, Sister Water, Brother Fire, and
Sister Mother Earth, drawing Sun and Moon and the four elements
into an expanded fraternal nexus and praising the Lord *through* each.
The question I would pose here is whether a democracy-to-come, an
expanded fraternity that broke free of the political, of the nation-
state, that perhaps points toward what in *Specters of Marx* he calls
a New International, might come to embrace the nonhuman inhabitants of the planet. They are clearly interested parties, stakeholders,
as we have come to say, in the fate of the earth, even if they have
no voice. As Christopher Stone argues with respect to trees,[33] this
is no more an impediment to their being represented in a court of
law—individually or collectively—than it is for other fictional legal
persons, such as corporations, or other voiceless persons, such as
infants and the mentally impaired. Of course, nothing would pre-

vent competing parties from claiming to represent the beetles or the elephants. But genuinely trying to represent the interests of a species, a region, or a watershed sets a standard that a court can take seriously,[34] even if we need to reimagine representation itself, given the fragile identification bridges by which we reach out to other species. This is not to say that there will not be countless impossible cases, aporias, dilemmas, and conflicts of interest. That is true of the earth itself—nature is as much a battle zone as a cooperative community.

The claim is not that a parliament of the living would bring an end to violence. It could only hope to manage an economy of violence.[35] But it might prevent or ameliorate what Derrida calls "the worst violence." He writes that "one must combat light with a certain other light, in order to avoid the worst violence, the violence of the night which precedes or represses discourse."[36] It is perhaps ironic that we are marshaling Derrida on the side of the voice when he has striven mightily against phonocentrism, the philosophical privilege of the voice. But the paradox is only apparent. For what becomes abundantly clear in the case of animal rights is that having a voice is *not* some original natural phenomena but essentially bound up with representation, just as it is politically for those humans of whom we say that they need to be given a voice. Is it intelligible to our thinking of living beings more generally as deserving of the ethical or political concern to "give voice to the voiceless"?[37] Aren't they voiceless in quite a different sense from humans who are being ignored, or from infants or impaired humans, who one day will, or once did, or might in principle have had a voice? It is hard to see why. Many creatures clearly do have voices—we simply cannot understand their calls and cries. And everything that lives has interests that can be met or frustrated. It is hard to doubt that being poisoned or losing the habitat necessary for survival is against the interest of whatever creature suffers this fate. Derrida's sense of a democracy-to-come, one that would break free from the constraints of the nation-state, one that would pursue justice beyond the rules laid down in advance, one that would take seriously the need to represent the interests of all earth's stakeholders could surely embrace this difficult but necessary ideal of a parliament of the living. It may be that its decisions would only be enforced once it becomes clear that, as Benjamin Franklin put it, if we do not hang together, we will surely hang separately. But even the thought of such a body begins to animate virtual voices in the human head, voices of creatures whose spectral presence haunts us in so many ways—species that have died out, flocks and herds we breed to eat, animal companions we live with, even the sports teams we name after animal totems. How long can we refuse to acknowledge all these ghosts, just as we balk at acknowledging the source and character of our own animating energies?

Conclusion

We began with two guiding questions: (1) Might not an econstruction—a living, developing, and materially informed deconstruction —find itself quite at home thinking through the quandaries of environmental concern? (2) Might not environmentalism provoke a certain materialistic mutation within deconstruction?

I have tried to argue this both ways—that environmentalism finds itself in an often problematic and aporetic space of posthumanistic displacement where deconstruction is particularly well equipped to offer guidance. Equally, environmental concerns can embolden deconstruction to embrace what I have called a strategic materialism, or the essential interruptibility of any and every idealization. Another way of understanding a deconstructive embrace of materialism would be this: we tend to think of matter and spirit or matter and mind as somehow opposed, and hence as unable to be thought of in the same space. But we can, I believe, get beyond this reductive understanding of opposition without falling into the arms of dialectical synthesis. The model of the Möbius strip allows for the idea that radical opposition can be combined with deep ontological continuity: at every point two sides but one surface. This gives us a beautiful way of representing man's relation to nature—opposed, in some sense, and yet at the same time continuous. Deconstruction allows us to think in this kind of space.

Finally, deconstruction's critique of presence leads effortlessly to the strange temporalities of environmentalism: we need to act long before we can be sure we need to, and it may only be our grandchildren who will benefit. The dangers we face are from the accumulated impacts of past practices. We are dealing with a singular sequence, the history of the earth, which takes to a whole new level the familiar idea that human history is ideographic—concrete, not rule governed, nor ever to be repeated.

I have further suggested that the problematic antihumanism of the '68 generation, far from being a dangerous ecofascism, is precisely adapted to our current situation, one in which the whole privilege of the human as a well-meaning but often toxic terrestrial is quite properly being put in question. I have argued for the renewed privilege of the human, by analogy with the privilege that Derrida conditionally accords Europe, as long as the new human can be understood as embodying a proper respect for otherness and difference. And problematic as it may be, I am suggesting we might extend Derrida's democracy-to-come to the (imaginary) parliament of the living. Derrida agreed that environmental destruction needed to be on any short list of the plagues of the New World Order. I hope I have begun to show how deconstruction as econstruction helps us address some of the complexities opened up by that extension.

CHAPTER

2

The Idea of Ecophenomenology

What *on earth* is ecophenomenology?[1] It is tempting to understand it as a "descriptive science," as Husserl characterized phenomenology. It might further be seen as the proper completion of the phenomenological tradition, with the work of Merleau-Ponty playing the role of a stepping-stone toward a fuller account of our sensuous earthly embodiment. This is an increasingly common suggestion.[2] Ecophenomenology could then be thought of as a hybrid: a joint venture between phenomenological ecology and ecological phenomenology. It would give an experiential grounding to the ecological vision and finally rescue phenomenology from any trace of idealism. Such an expansive approach would reaffirm the idea of phenomenology as a "rigorous science," committed to careful description. But the cool detachment of this very idea must be questioned. I argue instead that we need to acknowledge the essential ethical and political engagement of any ecophenomenology, even if these terms themselves are reshaped in the process. Such a reshaping would surely occur if ecophenomenology began to trace the shape of *any engagement we might have* with the world. I begin indirectly by trying to chart a way through the standoff between primitivist and constructivist views of nature, which belies such a fundamental engagement.

Philosophy of Nature

The philosophy of nature has a long and celebrated history, from Lucretius to German idealism and the present day. And yet there is a contemporary caution about using the word "nature" at all: it seems to reflect an outdated metaphysical naivety. Does the new upstart—ecophenomenology—dodge this fate? There is an inevitable overlap and interpenetration between ecophenomenology and the philosophy of nature. Over the past few decades—and in contemporary discussions—one *shared* concern has been to deepen critical reflection on the relation between "man" and "nature." What precautions must we humans take when reflecting on ourselves about (a) nature to which we oppose ourselves and yet belong?

In Husserl's hands, phenomenology toyed with idealism in attempting to step back from the constituted naiveties of the "natural attitude," to delineate the subjective activity responsible for this taken-for-granted world. Despite its best efforts, it had to go transcendental so as to ground a shared space of idealizations. The *natural* attitude was not that of nature but that of living habit. Nature, as such, was not available as a structuring resource. But while this return to a certain "activity" could be said to have shattered our habituated amnesia, it did so at the price of reactivating traditional metaphysical oppositions, such as those between subject and object, mind and matter, and nature and culture. Merleau-Ponty's development of an embodied phenomenology tries to overcome this problem.

At the same time, we have seen new, equally sterile standoffs between primitivists and constructivists about nature. Sterile but predictable, if understandable. For, on the one hand, we cannot do without the nature/culture distinction, and yet it is hard not to see the way it is cashed out as itself a cultural construction, albeit (as some would acknowledge) with massive material consequences.

At every turn, it seems we are entangled in questions of level. Bill McKibben's famous claim for the end of nature echoes—materially, as it were—the constructivist claim that the very concept of nature is a creature of culture.[3] McKibben asserts that there is no *pure* nature left, at least not on this planet, and that the effects of human agency are *pervasive*, especially through our effect on an all-pervasive atmosphere. But the absence of any pure nature is far from being an argument against nature. The weeds growing through the abandoned parking lot are not "pure," but they are surely a sign of a certain unflagging *resistance* to human paradise-paving schemes. Nature thought of as resistance *could* not be pure! Constructivism, too, harbors a puzzling blend of blindness and insight.

It is true that to communicate the idea of nature, one cannot just *point to a tree*, with what Hegel called sense certainty, focusing on this tree, here, now. Intuitions without concepts are blind, as Kant would say. But the converse is equally true—concepts without intuition are empty. The problem is construed much too simply if we work with a concept-object model in which a word, idea, or concept is linked to a more or less identifiable thing, such as "hand" and hand or "tree" and tree. The idea of nature is far more complex than a simple concept. First, as we have seen, while it is one concept among others, it is *also* a name for a whole category to which concepts are opposed, not to mention that which may well escape any and all conceptualization. Second, it is hard not to notice that "the idea of nature" serves all sorts of narrative and ideological functions, being lined up with virtuous, normal, "healthy," original, and so on. Torture, homosexuality, and zoos are condemned when they are described as unnatural.

What (implicitly) follows from this is that a critical engagement with the idea of nature *has to recognize* the articulated complexity of any/all of its conceptual manifestations, even as it inescapably points to something that antedates human existence, let alone our concept of it. Why do we say this? It is important to recognize that the idea of nature is complex, articulated, and multidimensional. And yet it is not that we each construct such concepts through a series of intentional acts (as traditional phenomenologists would suppose), but that we participate in such concepts in ways that unreflectively close down scrutiny over their ideological function. So far is it from being the case that nature is (just) a social construct; it is just because nature is so much not just a social construct that it matters *how* we (socially) construct it! Put another way, it is not a cultural choice that we "construct" nature in some way or another. We have to. For us humans, culture is a natural necessity, even if the shape we give it is not. And we can get it "wrong." Thinking of nature as pliant fodder for our "construction" may precisely lead to catastrophic climate change as nature laughs in our face. While the idea of nature does indeed function narratively, ideologically, in our culture (perhaps in all cultures), it does so precisely by being credited with a false unity and simplicity that a critical analysis can challenge, allowing it to be reassembled in different ways.

Let me give an example of how we can combine what I will call a critical construction with a provisional essentialism. *Any* adequate account of nature has to address such questions as: How do things move? Are there different kinds of change? What is life? Are things the way they appear? Do humans have a distinctive place in the world? Can the workings of the world (and/or its very existence) be explained in its own terms, or do we need to bring in other forces or principles? How do we distinguish between the physical forces that *drive* us and those (external) ones we face? How far can we *control* the forces "outside" us?

These questions are not ones we choose but ones we find ourselves faced with. What is *constituted* are the clusters of answers (concepts, narratives, maps) to these questions. If I have not provided a precise, exhaustive, and exclusive list of such questions, it is because our situation as mortal beings has a certain indeterminate, generative power. Our constructions of nature wrestle with such questions. But that they do so, that there are options, that some of our answers have outlived their usefulness may not be at all obvious. And perhaps the *best* way of shattering these illusions is to question the "we" (or "human") that recurs in most of these questions. What if this "we" were itself not homogenous but fractured? These options will reflect all sorts of other factors—level of technological development, patterns of social organization, relation to the land and agriculture, laws of property, religious outlooks, class and gender relations, and so on. Different concepts of nature often serve specific economic and symbolic purposes, disguising that fact by masquerading as neutral and ahistorical.

All this is to say that we can accept *and affirm* that the concepts of nature are contestable and indeed contested. Our own "culture" is the site of a struggle between a range of such concepts, each articulated as a set of answers to the kinds of questions we raised. But it is not *just* an academic struggle, and—to speak politically—the forces in play are not equal. Some version of the common diagnosis is hard to shake off—that the forces of capital have a stranglehold on the interpretation of nature as material for human consumption, whether or not theologically backed up with the language of dominion. If at one time, the technological orientation that opens up the world to our instrumental engagement was *liberating*, and perhaps it was, it is now commonly recognized to be the ideological equivalent of kudzu—smothering the more complex and sustainable relations to the world on which it was let loose, a plant that became an uncontrollable weed when its conditions of viability were breached.

The *intersection* of answers to these basic questions with normative and political issues seems broadly to occur through the expansion and contraction of questions of *scope*. Just as Kant diagnosed metaphysics as beginning when we apply the categories of experience beyond their proper sphere, something analogous can be said about the various *answers* to the questions about nature.

When control over nature takes the form of providing shelter against inclement weather, clothes for protection against the sun and rain, a regular food supply, and safety from predators and disease—who could not want control? And yet, we might say, the desire for control has gotten *out of control*. Local control is bought at the price of global instability, temporary control at the price of long-term unpredictability. Think of the production of drug-resistant bacteria through the use of antibiotics, habitat destruction through crop monoculture, allergy-prone bodies through overhygienic infancy, and so on. Where there is no ultimate "outside," externalization generates blowback.

The claim here is that the value of constructivism is tied to its potential for licensing emancipatory critique. This assumes, correctly I believe, that the need for emancipation is not just metaphysical (understanding aright our relation to nature), not just ethico-political (social and ecological justice), but also existential/material (concerning survival as a species). For such a constructivism is more than compatible with a robust sense of our inhabiting and indeed co-constituting a material world in relation to which we have to come up with answers to a range of fundamental questions. We need a quasi-Kantian insistence on the compatibility of transcendental constructivism and empirical naturalism.

We are, of course, left with the question of how to think about this floating set of fundamental questions, what connection they might have to ecophenomenology, and whether they are an ethically neutral ground for a subsequent ethics or are already steeped in the ethical and/or political.

Ecophenomenology?

One way of understanding ecophenomenology would be in terms of its expanded scope. Phenomenology would have evolved even in Husserl's own work through transcendental phenomenology, then the philosophy of the *Lebenswelt*, "the lifeworld of the 1930s," and now to ecophenomenology, gathering at each step further levels of resonance and contexts of significance. There is some parallel here with Derrida's claim, having worked through phenomenology, as he put it, that all meaning is contextual and no context is fully saturated. We would then be faced with the question of whether we were reconstituting a quasi-Hegelian progression, with the ecological taking the place of Spirit (*Geist*), whether the ecological was just a stage in a continuing progression itself to be superseded or whether the ecological could have woven into it a certain indeterminacy, a certain "ineliminable openness."

If we were to follow through on the Merleau-Pontean sensuous embodiment turn here,[4] the path to be followed by either the neo-Hegelian or deconstructive route would productively exploit the breakdown of the opposition between the intentional and the natural. Merleau-Ponty's word "flesh" is his way of marking this point. This breakdown can be encouraged from both ends by offering an evolutionary story of the developing complexity of intentionality and then encouraging the tendencies within "natural scientific" views of the world away from linear causality. On this latter path, the bleeding of the intentional into the natural world would take place at the level of fields of meaning and, for example, neural nets. But either way, we would be holding on to the idea of phenomenology as a science, a field of investigation whose structures we could, with luck, describe, even if it meant some fairly gymnastic peering into the place on which we are already standing. It would be an account of an essentially embedded (operative) intentionality, embedded by being embodied, with motor capacities, in a natural world whose substance we share.

What would it be to move away from phenomenology, or ecophenomenology, as a *Wissenschaft*, a science? Here we can learn something from Heidegger. Consider first his remark, about intentionality being the name of a problem (even as he was editing Husserl's work!).[5] Subsequently, in *Being and Time*, he sets intentionality within the structured matrix of fundamental ontology (Being-in-the-world). Then (in "Letter on Humanism") he writes of dwelling, and *ethos*, as the ground of any "ethics." Husserl's descriptive phenomenological practice gives way to an evocation of how it is grounded in the fundamental engagement of dwelling.

We could locate the leading edge of a certain ecophenomenology in something like this space of dwelling, a space that would coincide with that in which we ask and answer these fundamental questions about our place in nature. Why the leading edge? The claim is that there is a deep instability or tension about the line demarcating the descriptive from the ethical. To recap some of these questions: Are things the way they appear? How do things move? What is life? Are there different kinds of change? Do humans have a distinctive place in the world? Can the workings of the world be explained in their own terms? How do we distinguish between forces within and without? How far can we control the forces outside us? How do we come to terms with our own finite existences, our genetic and affective dependency on other humans, and so on? The claim is that these are some of the basic questions to which any philosophy of nature must give answers. Only then would we have a candidate for being a philosophy of nature.

Consider four of these questions.

1. Change. Should we respond by trying to control it? Or welcome and enjoy it?

2. Appearance/reality. This may be an old chestnut of an opposition, but we cannot do without occasionally wanting to step back to understand what is really going on, to probe beneath the surface.

3. Nature inside/outside. We need both to connect with our natural being and mark our separation from the "natural" (external) world in various ways.

4. Death. Imagining the breakdown of our capacity to imagine.[6]

Without being reductive, what is extraordinary is how much these questions take center stage in the broader history of philosophy, how intimately they are tied up with ethical and existential issues, and how quickly that dimension rises to the surface. What runs through all of these questions is the recognition of vital precarious relationality.[7] As soon as we open our ears and eyes, we are faced with a mass of changing information that we learn to make sense of in terms that matter to us in all sorts of direct and indirect ways.[8] We discover ourselves as in the world but not entirely of the world, with the capacity to stand back from it even as we cannot for a moment be separated from it, even as we are utterly dependent on it with every breath we take. We need desperately to hold its destructiveness and violence at bay, even as we explore the delights its shelter provides us. We must finally come to terms with the recognition that once we did not, and soon we will not, exist. These are not biological remarks; they are, perhaps, anthropological, in Heidegger's sense of *man* as a portal for ontological questions.

Suppose philosophy begins here, with these kinds of questions. We might then suppose that we need to answer them, as so many of them are tied up with anxiety, fear, and vulnerability—as well as joy and delight. Clearly, philosophy (and poetry and religion) has tried to do so. But so many of the answers have either generated further puzzles no less vexing (e.g., dualism), or they leave us cold. They leave us cold because they offer us systems, sciences, and grids that, however helpful they may be, require that we at least temporarily engage in a certain detachment from the world, which must then become our permanent residence, or the place from which we translate ourselves back.

If so, the ends of philosophy, and hence ecophenomenology to the extent that it seeks such a status, may not be best served by aping a descriptive science but rather by becoming a performative practice. I take my cue here from Nietzsche, from the clear intention of *Thus Spoke Zarathustra*, and from the obviously emancipatory ambitions of his narrative presentation and style, which he called a "dance."

The argument here, and there is an argument (one shared by Socrates, by Kierkegaard, and by deconstruction), is that the truth that philosophy discloses is fragile, evanescent, and unstable. We all too easily revert to simple oppositions, to blindness toward conditions of possibility, to staring at what is before our eyes. If phenomenology had a point, it was to return us to the *Sache selbst*—to the complexity of what presents itself to us in experience. Ecophenomenology would fulfill that promise by animating the natural and contextual conditions not only of what presents itself to us but to the possibility of that presenting. It is not enough that we attend to the phenomena; we need to learn better ways of talking about them and retaining their significance.

Three recent books offer ways of bringing a certain performativity of thinking to bear on the problem of the fragility of philosophical insight. The first book is *Being Alive* by Tim Ingold, the second *Vibrant Matter* by Jane Bennett, and the third *Against Ecological Sovereignty* by Mick Smith. Each, in their own way, models, performs, and encourages the kinds of relationality that ecophenomenology would seek to promote. In Heideggerian terms: expanded modes of dwelling.

Being Alive is subtitled "Essays on Movement, Knowledge and Description." Ingold is a social anthropologist who confesses to being an eccentric within his own discipline, drawing heavily on philosophers such as Whitehead, Heidegger, Merleau-Ponty, and Deleuze. This book continues a project begun in his *The Perception of the Environment*, best described as a conceptual reenvisioning and rethinking of our being in the world, where the central motif is that of being alive in the world. Its significance for ecophenomenology, though he does not himself use this term, lies in his explicit attempts at a thickening of our understanding of the constitutive relationality of our terrestrial dwellings, his various strategies for transforming our grasp of that relationality, and the ways we write about it. So, as with Whitehead, Heidegger, Deleuze, and Derrida, he embraces the idea that linguistic, even poetic, innovation may at times be called for, especially when existing grammatical formations reinforce misleading shapes of thought. Here, we might add, Nietzsche is an honorable but unacknowledged predecessor.

Ingold reworks something of the position that Heidegger explores in "Building Dwelling Thinking," suffused with an attention to perception drawn from Gibson, enhanced by Merleau-Ponty, and with a vital materialism drawn from Deleuze and Guattari. In the earlier book, he attempted to rescue anthropology (in the broadest sense) from metaphysically limited ways of understanding production, history, and dwelling, taking issue with Marx's commitment to the privilege of production and the supposed privilege of human history.

His basic argument is that to understand human life as centered on production should not just be corrected by doing dialectical justice to consumption. That retains and reinforces the model of action as actualizing an idea. Instead, we need what he calls an intransitive model of production as autopoiesis. "Producers, both human and non-human, do not so much transform the world, impressing their preconceived designs upon the material substrate of nature, as play their part from within, in the world's transformation of itself."[9]

Downplaying ideational anticipation has the effect of softening the
role of history as a self-conscious making by radically distinguishing the agency of humans and animals. Phenomenologists would see
here a parallel with embedding intentionality within the lifeworld,
as Husserl attempted in the '30s. What it opens up is a focus on engaged practices, both human and nonhuman, and linguistic formulations that assist us in articulating more complex relations between
subjectivity and objectivity, agents and patients, and so on.

Ingold's deconstruction of production and history continues in
his drawing upon Heidegger in understanding dwelling intransi-
tively, such that "building is not a means to dwelling" but rather
"a process of working with materials . . . bringing form into being."[10] More broadly, he speaks of attending more to "the material
flows and currents of sensory awareness within which both ideas
and things reciprocally take place."[11] He is very close here to what
Merleau-Ponty means by flesh (in *The Visible and the Invisible*).
And indeed, it is from Merleau-Ponty that he draws when distinguishing his position from that of Gibson, who importantly replaced
the Cartesian mind with an organism active in its environment.[12]
For Ingold, Gibson does not ask, as Merleau-Ponty does, the "transcendental" questions as to "what kind of involvement of the per-
ceiver in the lifeworld is necessary for there to be things in the environment to see."[13] Merleau-Ponty, in Ingold's view, ultimately sees
my perception of the world as the world's perception of itself.

Ingold moves on from reworking this triad of production, history,
and dwelling to mobilizing a neo-Deleuzian/Bergsonian view of
dwelling in the form of wayfaring, bringing in lines and movements
as central motifs somewhat as Derrida had worked with the trace.
Much of the argument rides on the ontological shift toward constitutive relationality, the primacy of process and change, and how
it affects the language in which we write. Consider just one of the
chapters in this book ("Landscape or Weather-World"), which looks
to bear on ecological issues.

Ingold sees in the very word "landscape" an optical prejudice
based on the confusion between "scape" and "scope." In a way that

echoes Aldo Leopold's preference for bogs and marshes over Sierra Club eco-porn, he both tries to wean us off the optical toward the haptic and have us step back from the optical by rethinking light and sky (in a way informed by Merleau-Ponty's "Eye and Mind").[14] Basically, he argues that sky is "not an object of perception" but "what we see *in*."[15] Again, he makes a transcendental relationality move: "We must imagine ourselves in the first place as participants, each immersed with the whole of our being in the currents of a world-in-formation: in the sunlight we see in, the rain we hear in and the wind we feel in."[16]

He goes on to describe the experience of walking on the beach in the mist, focusing on the weather, as a material form of what Heidegger called "the world worlding" and again, what he calls "sea-ing the land."[17] Instead of gazing out to sea from the solid land, we would conceptually and experientially bring the language of the sea—its waves, its surges, its turbulence—to bear on how we see the land.

Thales's response to the slave girl who laughed at the philosopher for falling into the well was to demonstrate his grasp of the cycles of time by buying up options on olive presses in a bad year. Ingold's plausible claim here is that language locks us into inattention, and in particular inattention to what I am calling our constitutive relationality, or the ways in which our very experience of the world echoes and draws on its essential dynamism and dissolves the illusions of detachment, separation, and objectification with which we surround ourselves. If ecophenomenology is to be more than an expanded "science," if the point of science (as it was for Husserl) is ultimately emancipatory, might we not need just such deconstructive (and poetic) turns of language?

Jane Bennett's *Vibrant Matters: A Political Ecology of Things* adds another dimension.[18] Political theorist Bennett attempts to explore what it would be like to attribute agency, or something like agency, not just to humans and nonhumans but also to material flows and formations, to a whole range of physical forces. One of her central lines of thought is to take the idea of intentional agency, to embody and embed it (with Merleau-Ponty), to disperse it to biological and social human capacities (with Diana Coole), and then to extend this initial materialism to cultural products (with Bruno Latour). The argument is plausible if not decisive. A parliament of things that extends way beyond the human is an imaginative provocation to us humans. And the explosion of the founding opposition between the natural and the intentional creatively works the same vein that moves us from phenomenology to ecophenomenology. This leads her to give articulation to atmospheres that, like Heideggerian mood, are hard to pin down.

Finally, Mick Smith's *Against Ecological Sovereignty.*[19] Smith is both a philosopher and a social theorist, and he has written a wide-ranging tour de force, integrating Agamben, Arendt, Nancy, Levinas, Heidegger, Latour, and others into a merciless exposure of the pervasiveness of the anthropological machine, the projection of human concerns onto the natural world, the presumption that it is there for us. It is not a book on ecophenomenology, but it allows us to pose our big question to it. We have pressed the idea that we may need to move beyond ecophenomenology as a descriptive science, because it exposes a constitutive relationality vulnerable to being covered over by our everyday habitual modes of thinking and is in need of critique. The truth calls for new ways of holding our attention, and, as we have seen, linguistic creativity is one way of bringing this about. Our fundamental engagement with the world, and co-constitution, could (just about) be understood epistemologically or metaphysically. But what if ecophenomenology did not just have ethical (and political) applications, but unavoidable ethical implications, or if it were already intrinsically ethical?

In a way, we have touched on this already, by referring to Heidegger's "Building Dwelling Thinking."[20] Dwelling is the place of ethos and the ground of any ethics.[21] At what point does an ecophenomenology touch on the ethical?

Mick Smith's *Against Ecological Sovereignty* connects the political critique of sovereign power with its ecological analogue—the reduction of nature to being a mere resource for human projects. In a way, Smith is reworking Heidegger's critique of technology and

Machenschaft, but with a more political edge and with a critique of biopolitics mediating the exercise of sovereign power over man and nature.

The question this raises for any ecophenomenology is this: Does not the question of sovereignty, of power, offer a very specific challenge to the idea of ecophenomenology (or, indeed, any phenomenology) as a science, rigorous or otherwise? The pathos with which Husserl writes of phenomenology as a science comes from his evocation of the shared pursuit of truth as one of humanity's highest values and his claim that only phenomenology will reconnect us to the fullness of our experience. Is the detachment and descriptive discipline of a science best equipped to achieve the reengagement that Husserl was calling for?

If Mick Smith is right, then just as we need to follow Merleau-Ponty in grasping the embeddedness of intentionality in embodiment, and embodiment as the chiasmic event of flesh, so we need to grasp the embeddedness of technology and calculation in the practices through which sovereign power is exercised. Echoing Philippe-Lacoue Labarthe,[22] we might say that ecophenomenology ends in politics. Of course, this is too simple, too reductive—because it also correlatively redefines and reforms politics. The important thing here is to recognize that we are no longer talking about an anthropocentric politics but one driven much more by a radical ecological perspective. It may be time to realize that ecophenomenology is, in the deepest way, a normative discipline, not for recommending this or that course of action but in articulating the shape and implications of what I have called constitutive relationality, both for humans and nonhumans alike.

Ecophenomenology would attempt to articulate this relationality in ways that would recognize the playing field of our habits and help us see, think, and write better. This points to a further commonality—between the descriptive, performative, and critical practice of ecophenomenology as philosophy, and the writings of David Abram, Gary Snyder, Aldo Leopold, or Gerald Manley Hopkins. Not to mention Bob Dylan.

It would be easy to understand such a claim in a hopelessly sentimental or romantic way—that poetry could save the world. There
is, however, a version of this claim—one anticipated by Heidegger—
that is less implausible, namely, that there really are a range of fundamentally different ways of being in the world, of managing our
relation to the world. Domination and cooperation perhaps capture
two of the most basic. Shifting from one to the other may be made
necessary by all sorts of things—war, economic hardship, the environmental crisis, and so on. But actually effecting these shifts works
through symbolically mediated practices. And there is no telling
what will prove most effective. This is the power of Heidegger's
speaking of preparing the way.

Changing the way we speak and write (see Ingold) really can re-
configure our practices. I will provide one last example, which very
much connects with Mick Smith's concerns about sovereignty. The
Woodrow Wilson Center published a document, *A National Strategic Narrative*, written by two senior military officers.[23] What is extraordinary about this proposal is the paradigm shift it proposes and
the language in which they capture this shift. Basically, the authors
argue for moving from containment to sustainability, from control
in a closed system to credible influence in an open system, and from
a defensive posture of exclusion to a proactive posture of engagement. They further argue for "acknowledgement of interdependen-
cies and converging interests" and "adaption to complex, dynamic
systems," and they talk of "strategic ecology."

Essentially, they are arguing precisely for a move from domination
to cooperation as a new national defense narrative. And the change
in language—what they call a new narrative—is essential. This language is not just descriptive; it also promotes and anticipates ways
of relating and negotiating, ways that reflect our actual interdependence. I am suggesting analogously that we set aside the idea of ecophenomenology as a science[24] and affirm it as an engaged practice,
one that bears witness to our fragile constitutive relationality.

3

Ecological Imagination: A Whiteheadian Exercise in Temporal Phronesis

Any physical object which by its influence deteriorates its environment, commits suicide.

—*Whitehead*

The significance of the temporal in nature has shifted dramatically
throughout the history of philosophy: Plato's grasp of the importance of the permanent, purposiveness as a source of significance
in time, time as a value immanent to historical change, and time
as tied to human subjectivity. With Whitehead, however (and later
with Deleuze and others), we have come to acknowledge something
new: the nonlinear multidimensionality and complexity of time.[1]
But it is one thing to acknowledge this complexity—it is quite another to bring it to bear on our thinking. This is especially needed in
environmental thinking.

Whitehead was not silent on the environment nor on the ethical aspect of how we understand it: "Organisms can create their
own environment.... [W]ith such cooperation ... the environment
has a plasticity which alters the whole ethical aspect of evolution."[2]
Grasping what might be called a dialectical or creative, interactive
relation to the environment gives us insights not available on a mechanical model in which the environment unilaterally determines
the response of an organism. But there is work to be done here:
"[T]he increased plasticity of the environment for mankind, resulting

from the advances in scientific technology, is being construed in . . . habits of thought which find their justification in the theory of a fixed environment."[3]

What we are calling a new temporal phronesis would be a major contribution to habits of thought, and indeed practice, that would not merely transcend the mechanical but move to another level the creative engagement with the environment that Whitehead calls for. To be out of touch, to fail to be attuned to the complexity of time is not just a cognitive but an ethical failure, where the ethical is to be understood not deontologically or instrumentally but in relation to ethos, or dwelling.[4] Such a phronesis would capture both the temporal complexity of the real, and of our engagement with it, and is central to the ecological imagination.

I focus here on this last stage, arguing that it is only by articulating and enacting such (often aporetic) complexity that we can achieve the attunement required for an environmentally sustainable ethos. This attunement is itself a creative accomplishment of the human organism, even as it sets limits to an unbridled sense of the transformability of nature.

For Whitehead, grasping and enacting such an understanding is itself an evolutionary development. As he writes "on the organic theory, the only endurances are structures of activity, and the structures are evolved."[5] His main focus, however, was on the creative possibilities of the human transformation of the environment. Philosophy would help: "pierc[e] the blindness of activity in respect to its transcendent function." Although in 1925, Whitehead was less taken by the tragic dimension of the human engagement with nature, it was not entirely absent: "Any physical object which by its influence deteriorates its environment, commits suicide."[6] Our temporal *phronesis* develops such an understanding of the creative as tied up with the precautionary. For example, as far as "nature itself" is concerned, we learn to identify tipping points, nonlinear processes, and transformative moments, and to think the multiplicity of interdependent but distinct temporal processes. As for our own agency and engagement in the natural world, we need to be alive to the complex temporal shapes involved in (1) restoring natural systems, (2) attending to natural cycles, (3) needing to act before we are sure we should, (4) timing (ripe and unripe times), (5) eventuation (enabling new systemic possibilities), (6) sustainable development, and (7) the perverse dialectics of action (avoiding reactive responses and regressive resistance to change).

But before developing the argument, I will first provide some background orientation about where I am coming from, and then lay out some of the respects in which Whitehead's philosophical position is fertile ground for these meditations.

Philosophically, my generation was moved by both Nietzsche's and Heidegger's attempts to think about dramatic possibilities for transformed futures, and the ease with which revolutions in thought end up reproducing the schemes they attempt to overturn. We were inoculated by French poststructuralism against taking Hegelian (or, indeed, any other teleological) understandings of history too seriously as well as the broad dangers inherent in an enlightenment project that we nonetheless cannot entirely abandon.[7] And some of us at least have been cautiously fascinated by Derrida's attempts to think the im-possible, the a-venir, and messianicity without messianism. Although it is not easy to draw the lines of demarcation here, in some ways Whitehead has more in common with the phenomenological tradition—including Merleau-Ponty and Heidegger—than with Derrida. More practically and politically, what draws our attention is the intransigence of the major problems of our time (e.g., terrorism, [im]migration, and global warming), their resistance to straightforward solution, and the irony and tragedy that has characterized our attempts to deal with them. In particular, global warming brings an unprecedented urgency to environmental questions, and, as much as Whitehead has to offer us, it surely raises questions for him, too.

Why Whitehead? I would single out the following points of reference:

1. Recognition of there being multiple temporal series, interdependent but distinct temporal processes. "The causal independence of contemporary occasions is the ground for the freedom in the universe."[8]

2. The interpenetration of subject and object as well as the centrality of affectivity/emotion/desire/concern in understanding their relation.

3. Repudiation of the mind/matter bifurcation.

4. Understanding of time in a concrete, "thick" sense as a (natural) process.

5. The possibility of creative freedom in the universe.

6. The essential relatedness of an organism to its environment and the different forms this can take.

7. The interpenetration of past, present, and future. "Take away the future and the present collapses."[9]

8. The role of human evolution in determining the shape of the future.

Here, I believe, we can graft an account of temporal phronesis onto his own analysis. In "Aspects of Freedom," Whitehead insists that thinking is not just a matter of reflective self-consciousness: "[W]e find ourselves thinking just as we find ourselves breathing and enjoying the sunset."[10] And instead of wheeling out Descartes as the villain, he will find support for this in Descartes's references to feeling.[11] He goes on to distinguish three inseparable levels or dimensions of thinking—instinct, intelligence, and wisdom. We are accustomed to valuing intelligence, but it is not enough: "The folly of intelligent people, clear-headed and narrow-minded, has precipitated many catastrophes."[12] If for the moment, we define intelligence as the ability to think within a frame, wisdom would consist in the ability to take into account (without calculating) the background conditions that the frame has necessarily excluded and to work with multiple overlapping frames. It is no accident that Aristotle's reference to phronesis in the *Nicomachean Ethics* is often translated in English as "practical wisdom." One question Whitehead raises later (in *The Function of Reason*) is quite how the word "practical" is qualifying "reason" here.[13] Might not the practical be the ultimate test of all reason (and wisdom)? We might be led to this conclusion by a prior suspicion of "abstract thinking," but equally, we might want to keep in mind that references to the practical can be useless in extreme situations and "impossible" choices. Might not wisdom precisely include, inter alia, the recognition of the limits of the practical, or that the practical can throw up aporetic situations? Formal rules may fail us, and rules of thumb, too.

If we were to rewrite Whitehead's distinction between instinct, intelligence, and wisdom in a way that more specifically applies to time, I suggest the following (which, somewhat ironically, I present in a quasi-teleological timeline): However we describe what precedes the advent of life on earth, we may imagine the most primitive life as being "trapped" in the moment—certainly as having no consciousness of anything else. Later, within what we call instinct, we can discern engagement in temporally extended activity, some sort of unreflective awareness of possibilities, and hence some significantly temporal awareness, even if it is not an awareness, as we would say "of time." Taking a step further, we find what I will call "planning," which includes induction and empirical calculation of natural regularities as well as cultural projection—setting up the calendars, timetables, and artificial orders by which we *make* life predictable.[14] Of course, we often find mixed modes where cultural units are derivative from natural ones—the *day* being the most obvious example. We rightly admire the extraordinary astronomical achievements of earlier civilizations that allowed them real management of agricultural cycles and corresponding symbolic activity. In Whitehead's terms, this takes us as far as intelligence. But it is at the next stage that phronesis kicks in.

Phronesis as practical wisdom in its temporal specification is already familiar enough to us in countless proverbs. Be prepared. Cross that bridge when you come to it. Forgive but don't forget. Live for today, for tomorrow never comes. A bird in the hand is worth two in the bush. All things come to he who waits. Interestingly, such proverbs often point in different directions, advocating, for example, both enjoying what is available now and having the patience to wait until tomorrow. This suggests an important truth that these proverbs at least do not help us with—that we can expect to find ourselves in situations for which we have no definitive guidance—not only no clear rules but no unequivocal rough maxims, either. Here we approach the terrain of what I am calling temporal phronesis.

It is not that there are no rules; it is that we need to remind ourselves that there are many, and that the conditions and circumstances in which they are appropriately applied are not themselves subject to rules. Whitehead's claim—justified or not—that efficient causation is not enough to explain the cosmos, and that we need to posit final causation and creativity, would be an example of this claim. Many other examples cluster around the limits of the future's representability, and we are not entirely abandoned by proverbs here! "The best-laid plans of mice and men may go awry" nicely captures not just unpredictability but uncontrollability, the way in which, as Sartre put it, "Les choses sont contre nous," the way the real resists our will. When we think of the environmental crisis that stalks us, we find something of the converse situation—that a change with a real shape to it (global warming, mass species extinction) is happening behind our backs—the unintended aggregate outcome of a vast number of human acts of varying levels of intendedness. Why, then, do things happen that we do not predict? The obvious answer is that we lack information, or that we have an inadequate model of what is happening, or that we have our collective head in the sand. There are fields of concern in which we do get things right. If I cook a meal from traditional ingredients in a customary way, I would be rightly surprised if it turned out to be dreadfully inedible. The fact that my neighbor across the street played a CD of a new recording of Verdi while he prepared the same meal should have no effect on the cooking. The culinary factors relevant for success seem to be under my control. There are clearly relatively autonomous causal spaces in which what was true yesterday is true today, and much of what we do successfully operates within such spaces. But human history is not one of those spaces nor is the history of the planet. History, as Windelband put it, is ideographic, not nomothetic; it is a singular event, not one that will or could ever be repeated. This makes it a tough case for empirical prediction because there is no class of sufficiently similar events from which one could inductively generalize. Environmental events, one might say, are notorious for eluding prediction for a curious but essential reason, namely, that the "environment" is precisely what typically functions as background to phenomena that are being thematized. Background phenomena resist the lawlike idealizations from which scientific laws can be drawn. The planet as a whole compounds this problem by aggregating all the backgrounds without limit. Prediction is only possible when we can isolate overriding trends, like the buildup of CO_2. And then we can speculate about the consequences of this. For many of us, watching this happen is a variant of Kant's sublime, as when one watches a storm rage from the safety of a window, until the day comes when one wonders whether the window will hold.

The general argument here is that temporal phronesis begins at the limits of our representation of time, especially the future. It concerns itself with the habits of thought appropriate to situations of unredeemable complexity, that is, situations that cannot usefully be reduced to simple models for which we can generate rules. There are analogues here to Derrida's claim that a responsible action requires that we pass through what he calls the undecidable, that is, the recognition that we have no decision procedure, no algorithm, to lean on.[15]

However, it would be a mistake to suppose that we are somehow abandoned to struggle in a raging sea beyond representation without any help or guidance. I will now suggest some of the conceptual resources that may shed light on our task. My aim here is not to exhaust the field but to illustrate, however provisionally, what I mean by temporal phronesis. I distinguish between significant temporal features of natural processes and considerations relevant to our response to them.

What I am calling temporal phronesis is not exactly new. Identifying it in this way may itself be an exercise in temporal phronesis—drawing from the tradition the resources for a postlinear thinking about time, so as to connect with our own temporal engagements. Implicit in this account is the suggestion that temporal phronesis is something of a concretizing corrective to unidirectional and unidimensional models that have only ever worked under artificial conditions but which have acquired an undeserved authority.[16]

If I were to draw up a list of significant temporal features of natural processes, a list that dips in and out of considerations that Whitehead would highlight, it would include:

Natural cycles
The multiplicity of semiautonomous temporal series (founded on
    relational relevance)[17]
Entropy, both local and global
Sustainable development
Nachträglichkeit and regressions of all sorts
Anticipations of the future[18]
Patterns of change
Sports of nature, chance, creativity, novelty
Irreversibility
Negentropy, self-organization, autopoiesis, self-maintenance, and
    purposiveness
Discontinuity, accelerated change, tipping points, transformative
    moments, nonlinear processes, endings, beginnings
Identity and repetition

I will selectively comment on the last four of these items: irreversibility, negentropy, discontinuity, and identity/repetition. First irreversibility. In a sense, everything is irreversible. And yet so much of our ordinary behavior presupposes otherwise. Mistakes can be rectified, insults apologized for, damage fixed. But what I mean here is that we have devised many ways to generate technological and/or symbolic equivalents that can give the illusion of reversibility. The massive epidemic of species loss, however, gives the lie to this charade. What is tragic far beyond calculation is that what is lost in each species is not just a past but a future, a creative transformative possibility that took eons to evolve and whose like will never be seen again. Negentropy is another name for what Whitehead sees as the creative force in the universe that counteracts what would otherwise be its inexorable and dispiriting decay. Negentropy clearly interrupts linear models of efficient causality, and it raises the question—physical, existential, even theological—of whether it enters the cosmos only when life appears or whether it pervades it from the start. And it opens up the broader space of holistic explanation and puzzling questions about explanatory adequacy—such as are associated with Lovelock's Gaia hypothesis. Tipping points and discontinuities again ruin our linear understandings of things. More importantly, to the extent that the change we most fear is unprecedented, they destroy a certain assurance of calculability with respect to the future. This *could* be a good thing, forcing the adoption of the precautionary principle, where that is not itself problematic. (We think we *know* that such a principle is applicable to giving up smoking, to cutting greenhouse gas emission. Temporal phronesis, as an orientation alert to the counterintuitive and the perverse, might at times suggest caution about caution.) Finally, identity and repetition: for Whitehead, dull repetition ("fatigue") can signal the death of the upward movement of the cosmos ("relapse into the well-attested habit of mere life."[19]) But we also recognize that repetition is the form through which identity is temporalized, that living (and cultural) things have to repeat and be repeated to survive, and even need to be transformed. *Plus ça change, plus reste la même.* This helps us resist the temptation to resist change *in the very name of identity*. It may be its very condition; I think it is what Whitehead calls adventure.

To these complex temporal features of natural processes, ones that, broadly speaking, temporal phronesis would call on us to attend, I add a list of modes of orientation that similarly illustrate what I mean by temporal phronesis.

That something like temporal phronesis is needed is made plain by the story of the boiled frog. The problem was not that the water temperature was slowly raised but that the frog did nothing about it until it was too late. Perhaps the germinal form of temporal phronesis would be to expect the unexpected. But this must not too quickly reduce complexity to being a new object of old-style expectation. Something of this is important, but by analogy with Heidegger's discussion of being-toward-death, the very shape of expectation may change, and Whitehead points us in this same direction.

Expanding our sense of a transformed orientation to time, we need to bear in mind the following considerations:

1. The course of time presents opportunities for intervention, what the Chinese might speak of as ripe and unripe times; it can be too soon or too late. This issue becomes centrally entangled with questions about knowledge and evidence when we realize that the evidence that would, under a certain model, justify intervening might not be available until it is too late. Many fear we are in just such a position now with respect to global warming. Something like the precautionary principle is forced upon us by this thought.

2. We are also in a position to bring about, to welcome, to solicit, and to encourage creativity, invention, and eventuation—enabling new wholes, new spaces of possibility. This involves understanding creativity not as extended control but as enabling what will exceed my control.

3. In the environmental sphere, we are quickly faced with the problems and paradoxes of restoration, and how to justify "returning" to this or that supposedly original or privileged position.

4. We need to learn (and relearn) the dangers of generating what I would call reactive responses and stiff-necked resistance. Recent military adventurism would suggest that Temporal Phronesis 101 would involve a crash course in the dialectics of action. Much intervention produces just the opposite of what one intended. Ecologically, this has frequently happened with the introduction of new species to control pests that then turn out to be pests themselves (kudzu, rabbits, honeysuckle, Mexican feathergrass).[20]

5. A certain model of restoration exemplifies a wider shape of desire—to suppose there is an ideal state of rest to which we can return, toward which we should aim, what Heidegger and Derrida would call the lure of "presence." As a process theorist, Whitehead explicitly repudiates the idea of a final resting place. "The object of this discipline [speculative reason] is not stability but progress." I pause, however, when Whitehead writes of the coordinated nature of things as "an infinite ideal never to be attained by the bounded intelligence of man."[21] To my ear, this has unfortunate echoes of Hegel's absolute spirit.

6. In general, Whitehead resists the fixity of any distinction between subject and object in favor of their coordination. And yet he clearly does allow, and we need to emphasize, the consequences of what might be dubbed fallen, inadequate forms of consciousness. The fallacy of misplaced concreteness is not just a philosopher's error but an often tragic condition. And when (as we saw earlier) Whitehead writes that "the folly of intelligent people, clear-headed and narrow-minded, has precipitated many catastrophes,"[22] he is saying that metaphysical misunderstandings have had real-world consequences. I conclude from this that temporal phronesis does not just supply productively complex modes of orientation; it also has a critical, demystifying role, exposing illusion.

While certain theological motifs in Whitehead are worrying, he is right that we can and should be considering not merely more complex counterintuitive shapes of time but also a certain transformation of our general grasp of temporality. Here I would link Whitehead's account of the immanence of eternal objects, and of future ideal possibilities in the present, with Heidegger's discussion (in *Contributions*) of the role of the kind of thinking that would succeed philosophy as that of *preparing the way* for the reappearance of the gods, keeping alive, we might say, certain possibilities. And I would connect it to Derrida's promotion of what he calls the impossible, or what we might call unthinkable possibilities, possibilities "outside

the box." If we cannot predict or expect these, we can perhaps anticipate them—keeping a place at the table for such a stranger even
if he never turns up, keeping alive the candles of peace, justice, and
freedom not just as hope but as real possibilities whose conditions
of realization have not yet arrived. Here we would be echoing Che
Guevara's words: Be realistic, demand the impossible. The argument
for revolution is not difficult to make once it is clear that business
as usual cannot work.

Whitehead's view of the cosmos as a creative activity, and some
of the metaphysics that goes with this view, is not wholly implausible. If I read him correctly in "The Function of Reason," however,
we will part company over whether this creativity is, from the beginning, evidence of final causation. For life is arguably a prerequi-
site for finality, even if it is preceded by novelty and increasingly
complex structures. He appears to believe that evolution is driven
by final causes, while I believe that final causation is a product of
evolution and is restricted to living organisms.

His argument for cosmic final causation is flawed. The fact that
we can find purposiveness in human and other forms of life tells us
nothing about what operated prior to that. Complex arrangements/
distributions of matter can occur unintentionally, but having occurred, they can come to possess stability. Shake rocks in a jar, and
they will settle into a shape occupying a compact space from which
it will be hard to dislodge them. (Compare building a stone wall.)
What, then, is life? Suppose we identify as its key characteristic the
capacity of a certain organized complexity of matter to maintain itself and/or to continue to develop something of similar complexity.
We may ask: How could this happen without there being some will
or drive or force urging life into being? But this is simply a temporal
illusion. What looks like creative directedness is surely the mere
consequence of success. Self-replication, self-maintenance, is think-
able as a product of the random arrangements of matter. Suppose
the by-product of the rusting action of water on iron is water, which
then rusts more iron. Imagine this creates holes in iron ore deposits
so that more iron collapses into these voids, to be "eaten." If such

patterns can emerge by chance, they are sustained not by chance but by the fact that they are self-replicating. Such patterns can arise prior to what we call life; life arises when there is a chance conjunction or association of such patterns in which there arises an ability to bend the environment—by both positive and negative feedback—to the sustaining needs of the new organism. In such a way, purpose arises in the cosmos. Efficient causality is not simply linear and law-governed in the sense that the law already exists. For efficient causes occasion possibilities of complex coordinated relation that are wholly new. That a new self-organizing system arises does not require finality; it only requires an efficacious conjunction of parts and subsystems, which together make a certain replicating process possible. Whitehead seems to reject this position. In "The Function of Reason," he plausibly insists on the impossibility of thinking of human and animal life in mechanical terms. He properly concludes that the universe cannot be wholly thought in this way. But then he makes the leap that "[t]he material universe has contained in itself, and perhaps still contains, some mysterious impulse for its energy to run upwards." And that "the operations of Reason constitute the vast diffused counter-agency by which the material cosmos comes into being."[23] I suggest that it is precisely at another level of temporal phronesis that we can see that such claims are retrospective illusions. A name for this might be the consequent nature of teleology. On the ratchet model of history, there is a tendency for any arrangement that can preserve itself to do so, and for self-sustaining beings to sustain themselves. If these sound like empty claims, they are not. They explain how it is that even by chance, and even in the midst of a general tendency to entropy, later phases of the cosmos are increasingly populated by things exhibiting stability and self-maintenance. Once they appear, they tend to stick around. By definition. To see this, it helps to engage in a little separation of subject and object. Affectivity may not be eliminable, but that does not license any and every metaphysical desire. In short, the energy of the universe may "run upwards," but we do not need a "mysterious impulse" to explain it.

Whitehead is right to think that the creative process really comes into its own with life, and in particular with human beings. There is, however, a certain naive optimism that he himself might have questioned had he lived to engage with our current environmental crisis.

We began quoting remarks he directed against understanding evolution as adaptation: "Organisms can create their own environment. . . . [W]ith such cooperation . . . the environment has a plasticity which alters the whole ethical aspect of evolution."[24] And we read, too, that "[a]ny physical object which by its influence deteriorates its environment, commits suicide."[25] But Whitehead does not take the ecological turn that sees the deep shape of our relation to the environment to be toxic. He speaks of our threefold urge—to live, to live well, and to live better.[26] He continues, "[T]he primary function of reason is the direction of the attack on the environment." The only note of caution here attaches to those "races" "goaded onwards, sometimes to their destruction" by "that touch of the infinite."[27]

Our account of temporal phronesis reflects a somewhat more chastened view of human creativity, drawing on Whitehead's original and productive understanding of natural process, time, and creativity. This account, only anticipated here in bare outline, would itself be a creative evolutionary accomplishment of the human species. It would take further the goal of what he calls speculation, of developing "a practical technique for well-attested ends." Important as reflection is, its fruits need to be turned into both new habits and transformative practices. My view, for what it's worth, is that the complexity of our human temporal engagement with the world is not itself new. But our current climate catastrophe forces us to take it seriously rather than relying on simpler models, just as seafaring and then air travel made flat-earth theories obsolete, even if they still safely get us downtown and back. If, as I suspect, we are talking about a revolution, including new patterns, rules, dispositions, formulae, and coercive and persuasive habits, the content of such a revolution will not be limited to temporal phronesis. It may be that the frame for this new conjunction will have to be a new sense of spirituality, one that addresses the practical implementation of ultimate ends. If this sounds too New Agey, or too much like "Only a God Can Save Us Now" (Heidegger), reference to a "new spirituality" could be taken more simply as an *economy* through which self and other, man and nature, are distributed in a sustainable way. An economy one could affirm and live by.

4

The Eleventh Plague:
Thinking Ecologically after Derrida

The American way of life is not up for negotiation. Period.
—*George H. W. Bush*

The Deconstructive Disposition

What is it to inherit the work, the writings, the insights of another? This is not unconnected to "What is it to read?" The problem that arises here is not unfamiliar to admirers of Nietzsche, who cautions against those who would be his followers, *Folgende*, the mathematical expression for zeros at the end of a number. Are we then to follow Nietzsche's proscription against following him? Derrida has been condemned by some ("Dogs bark at what they do not understand" [Heraclitus]) and drawn into empty culture wars by others. He himself hardly ever tried to correct or contain this profligacy. But all those of us who followed Derrida and learned from him at some point or another face the question of inheritance. Derrida animates the question of inheritance in *Specters of Marx*, offering a model that would require selection and creative transformation. Moreover, as he insists, a gift sometimes calls for ingratitude. At what level can or should we apply these ideas to reading Derrida himself? Do we have to transform the idea of transformation to avoid just following him? Or would that not be the most faithful, and hence least faithful, response? To be faithful to Derrida, do we have to betray him?

80

Derrida might not endorse this language, but I propose here a re-working of Heidegger's account of what it is to engage with the work of a great thinker—he speaks of not going counter to the other but going to their encounter (with Being).[1] To do this, we have to bring our existing passions and commitments to the table. He is saying that we have to address what is most at stake in the other's thinking and writing. Moreover, we have to have skin in the game. In glossing Heidegger's claim in this way, I am bypassing the problem of endorsing his specific thesis about the primacy of the question of Being. A less technical cousin of this claim has legs independently of Heidegger's specific formulation.

Analogously, thinking about how we can *follow* Derrida without falling into aporetic elephant traps, we can draw on a distinction between doing as he does because he says so, or does so, and doing as he does because it is a smart thing to do. Put less casually, Derrida's ruminations about (re-)reading Marx are themselves not completely original, and no worse for that. Context, and the space of concern, changes. Marx was not saying just one thing but drawing together multiple threads from which we, in our time, cannot but select. Licensing ourselves to do this can enhance the transformative creativity of our response. It is in this spirit that I advance the idea of a deconstructive disposition. And in response to the ten plagues that Derrida names in *Specters of Marx*, I insist on an eleventh plague—our growing global climate crisis.[2] To honor Heidegger's formulation at the same time, it would be necessary to construe this reference to an eleventh plague at something like an ontological level, without being caught up in the seductions of "ontology." Forging an amalgam from Derrida and Heidegger, we would try to show that the eleventh plague was not just "one more plague" but was at the heart of the first ten, or at least was intimately implied or caught up in them. In the most summary form, this would be to show that questions of violence, law, and social justice are inseparable from ecological sustainability. A similar move would demonstrate that another candidate for the eleventh plague—the animal holocaust—is closely connected both with the first ten plagues and ecological sustainability, perhaps serving as a bridge of sorts. I will only gesture at such an account here.[3]

What is meant by a deconstructive disposition? The danger of such an account is that it may seem to dilute what deconstruction has to offer by blurring how it differs from other modes of "critical" reading. I will address this shortly.

I propose four dimensions to a deconstructive disposition.

1. Negative Capability

Keats described this as a willingness to tolerate ambiguity and uncertainty.[4] This is not to license intellectual laziness but rather to caution against premature closure. Derrida's reference to going through the undecidable could be understood in this way. And, indeed, the broader willingness, even passion, to disturb the sleeping dogs of (often binary) complacency.

2. Patient Reading

The point of reading (and thinking) is not simply to understand, using the handrail of existing meaning, but to open up possibilities. This requires patience, even when we have no time! What does such patience yield? It allows us to restore repressed differences and to expose invisible framings and stagings, even of the very occasions at which issues are being discussed (see Derrida's prefatory remarks to many of his presentations, raising such questions as "What is an *international* conference?").

3. Aporetic Schematization

Thinking often takes an essentially aporetic shape: the past that was never present, the gift that resists gratitude, a supplement to what is already complete, the always already, forgiving the unforgivable. These shapes need to be exposed and worked through if only to grasp the complex underbelly of intelligibility and coherence.

4. Attention to Language and Terminological Intervention

Language is not neutral. Words harbor ways of seeing and being in the world that are sometimes limiting or regressive. We can intervene in this invisible process with careful attention to these frames and by actively bending old words and inventing new ones.

It might be said that none of these dispositions are exclusive to deconstruction. So is there not a danger of dilution? Deconstruction in the late '60s and early '70s was an *event*, an interruption, a challenge, one attuned to the intellectual scene of the time—structuralism, semiology, a quiescent Marxism, pervasive doubts about the complacencies of humanism, of psychoanalysis, of phenomenology, and of literary theory. (I am trying to cover here the French and Anglo-American situations, which were different.) But it is no longer an event. Its covert influence has waned, even as the scholarly industry prospers, and some of those strongly influenced by Derrida delight us with their own brilliance and originality. Moreover, it is not entirely a bad thing that deconstruction should have metastasized in many directions, even if its pedigree is less visible. But deconstruction would cease to be deconstruction if it became an idol, an orthodoxy, a citadel to be defended. It can live on only as an event of renewal, a *repetition* of deconstructive strategies, gestures, and sentiments in the context of a new urgency.

Calculating and naming the inheritance of deconstruction is a thankless and unending task. The most salient threads that specifically address environmental concerns would include: language, time, the animal, sovereignty, topological complexity, the new international, inheritance, and death.

I propose to make some remarks here about the first three of these threads.

Unsettling Language

Deconstruction's bad press began with the phrase "There is nothing outside the text," which sounded like linguistic idealism. It was

later reworked as the ineliminability of con-text and the impossibility of ever completely specifying that context. But language itself continued to ground both hesitation and creative response. The
normative commitments of words such as "parasite," "rogue" (nation), "proper," and "authentic" all rest on structures of asymmetrical binary privilege that can be exposed and perhaps destabilized by
inserting an *indecidable*. And as Derrida showed in "Des Tours de
Babel," translation, which highlights the instability of proper meaning, is a powerful site for deconstructive archaeological excavation.

Consider three classic examples:

1. It is said that the bombing of Hiroshima was ordered after Prime
Minister Kantaro Suzuki responded to the demand that the Japa-
nese surrender by using the word "*Mokusatsu*," which can mean to
"ignore/not pay attention to" or to "refrain from any comment."
The former could reasonably be considered a refusal to surrender.
The latter was asking for more time. Could the subsequent loss of
nearly 250,000 lives in Hiroshima and Nagasaki be put down to a
mistranslation of a nuance of meaning?

2. The cult of Mary, the miraculous character of Jesus's birth, rests
on the translation of the original Hebrew *almah* (הָמְלַע) as "virgin,"
when it more accurately meant "maiden" or "young woman."

3. The license given by the translation of *rada* as "dominion" in
Genesis 1:16/1:26, as God's understanding of the relation between
man and the other animals and nature more generally, has been argued to be the source of much Western complacency over the destructive and exploitative consequences of man's reign over nature.[5]
Some have argued that it should be translated as "hold sway" and
others as "rule," with the strong implication of the benign responsibility that might be expected of a thoughtful ruler. Others have
pointed out that the literal meaning of *rada* is "a point higher up
on the root of a plant." Such a point is where the strength of the
plant as a whole is centered, offering a more collaborative sense of
privilege.

This last example has direct relevance to ecodeconstruction.[6] The authority of canonical texts is a continuing issue, considering the continuing reverence accorded to the Bible, the Koran, and other religious writings. This would be true even if there were no issues of translation. But the hermeneutic mischief with which they can be treated seems to know no limits. Translation is often a political act, calculated to inscribe or reinscribe power asymmetries. The demonstration that even an authoritative text contains within it competing meanings and possibilities allows other ways of reading it to be opened up. And, of course, this applies not just to the Bible but to the US Constitution, to the Pre-Socratics, to Aristotle, Kant, and so on. Latour's recent treatment of Lovelock's Gaia hypothesis, in which he attempts to empty it of any residue of political theology, is a good example of how the slant of such readings can make a difference to environmental thinking. Gaia can be ridiculed as New Ageism or (with Latour) mined for its progressive potential.[7]

Central, too, to any ecodeconstruction is the force and meaning of the word "nature," caught up, as it is, in binary opposition to culture, to man, to spirit and functioning repeatedly as a transcendental signified, a ground of meaning that would escape the play of language and the very oppositions in which it is inscribed. We are all acquainted with nature, whether it be last year's tornadoes or this year's tomatoes. Nature seems, straightforwardly, to be what's "out there," something we realize we are part of when we feel hungry or get lashed by heavy rain. But "nature" is not just what is real, what is out there. When placed in opposition to "culture," it has played a powerful cognitive role in organizing human life and thought. One of the hallmarks of early deconstruction was to problematize this simple opposition. It is clear, for example, that we approach nature through all kinds of cultural mediations and constructions, which themselves change throughout history. And these cultural constructions are not just shaping or distorting lenses; they often lead directly to transformations of nature. (When "nature" is treated as a resource, a mountain becomes a pile of quarry stone.)

Ecodeconstruction reflects our hope that we can get clearer about
the complex role that "nature" plays in our thinking, in our understanding of ourselves, and in our practical existence. This issue
is important in academic life, not least because universities are
constructed on the basis of distinctions between natural sciences,
social sciences, and humanities, as if these were separate fields of
inquiry, distinctions that depend on how we think about nature.
Deconstruction has made it more normal to inspect the boundaries,
the frontiers, the contaminations, the difficulties in making these
clear-cut distinctions.

While there are those for whom this distancing (from a naive sense
of nature) comes easily,[8] there are others who resist, who recognize
the desire to point and say, "That's nature," that striving, pulsing
force that precisely escapes description, like Roquentin's black root
in Sartre's *Nausea*. The question we are left with is this: Is it possible to accept that any concept we have of nature, any meaning we
give that word, is culturally constructed, riddled with narrative, and
as such burdened while insisting that there *is* something we are in
different ways culturally constructing? Much interesting work has
been done critiquing the idea of a return to some original "natural
condition," the restoration of a pristine origin, the protection of the
purity of wilderness. As Bill McKibben wrote long ago (see Chapter
1, this volume), there is no nature anymore.[9] Nothing with air blow-
ing through it has escaped human influence.

So the analysis is taking place at two different levels. The concept
(or sign) of nature cannot escape the cultural conditions of concep-
tuality. And "nature itself," materially, has been contaminated by
human activity, destroying the purity by which it could function in
opposition to man.[10] All this presupposing that man was not always
already part of nature. Is this instability regarding "nature"—that so
often simultaneously includes and excludes "man"—a distinctively
Western thought?

Arguably, our dominant modes of engagement with the natural
world are the reflection of narratives, often what Jean-François Lyotard would call grand narratives, such as man's God-given sover-

eignty over the natural world, or man's place in the great chain of being, or the story of enlightenment, in which inferior races, religions, and cultures suffer the same subordinating fate as nonhuman creatures, a fate in which these various disparagements are often roped together.[11] This presents us with an option—either to abandon the whole grand narrative scene in favor of multiple local, smaller-scale narratives or to continue with narrativity as indispensable while interrupting it. Or replacing an oppressive grand narrative with one with more of a future. Recall that Derrida wrote of the Bin Laden narrative: it "does not open a future."[12]

The upshot of these debates is itself a contested space. Some would use the constructedness of "nature" as an argument against any critique of technology that would accuse it of sullying our natural condition. Others more reasonably argue that we need criteria other than protecting or restoring purity by which to evaluate our engagement with the earth. These considerations all develop from reflection on the word "nature" and the *desire* attached to it.

It is not always clear whether these issues are linguistic, conceptual, or empirical, but the scope we give to words such as "pain," "consciousness," and "person" has a direct impact on how we engage with the natural world. Allowing that animals feel pain qualifies them for our consideration. Crediting them with consciousness bestows further rights. Indigenous peoples often attributed personhood to nonhumans, which strongly shaped their engagement with them. We might conclude that anthropomorphism is a precondition for moral consideration, but surely that begs the question. A broader sense of biomorphic personhood works just as well. Of course the scope of personhood has major consequences across the board from the biopolitics of abortion, to the rights of corporations to recycle their profits in such a way as to promote climate change denial through the corruption of political discourse by lies and sophistry. It is as legal "persons" that America protects the freedom of billionaires and multinationals to "speak" with their wallets.

The issue of whether we are indeed dealing with climate change or global warming is itself contested. Those who deny it tend to call it climate change, while those who accept it and its anthropogenic cause, and want to do something about it tend to speak of global warming. Moreover, it is surely remarkable that we do not have much of a name for what is likely in store for us, which is climate catastrophe. Here I have in mind Derrida's provocative remarks about 9/11, where he argues that our use of a date to name this event reflects an inability to grasp what actually happened, much like a traumatic event.[13] The real event, he suggests, is not what happened on the day but the hole it blew in our sense of security, making us wonder, "What next?"

The matter of the name is of deep significance in thinking about global catastrophe. As Lacan showed in his discussion of the symbolic phase, having a name is an ambivalent phenomenon. On the one hand, it functions as a handle by which others can manage us. On the other hand, it is a source of self-integration, recognition, rights, and so on. In the light of this, it is of particular interest that the mass species extinction currently under way is taking place with most of the more than eight million species not even being identified or named. Their very existence is a statistical extrapolation. If there is something tragic about threats to species we know (e.g., the snail darter fish in the Tellico Dam on the Little Tennessee River in 1973), there is something beyond tragic in the extinction of species we humans have never even identified as such. Nietzsche laughs at philosophers' supposed concern with truth, telling us that, like every other creature, we are really only concerned with what contributes vitally to our lives.[14] And yet at another level, we are deeply committed to knowing what is going on around us, and, reflectively at least, anonymous extinction is surely a shameful matter. Joni Mitchell's lyrics "You don't know what you've got 'til it's gone" is the optimistic version. More truthfully, you still don't know—we will never know—even when it's gone. The power of the name is real. The masses of animals industrially slaughtered for food die without names, many even without numbers. Their individuality is preeclipsed as mere stuff. Even chicken #2013783456 is a source of chicken stuff. A name is no guarantee of protection. At times, it can be a death sentence.[15] But it does draw you in to the symbolic and to the possibility of negotiation. Similar issues are raised by concern for future generations of humans, who as of yet have no names, and indeed do not exist, and yet arguably have interests that need to be taken into account.

When Derrida talks about animals, three obvious points stand out. First, he claims that the very word "animal" is pretty much a license to eat, or at least to consume in whatever way suits us. It has little, if any, biological significance, occluding every difference between the creatures it subsumes. And instead of talking, he presses the point by coining the word "animot," highlighting the way language has here been captured by the anthropological machine. Finally, consider his willingness to speak of animal genocide and link their fate with the holocaust in the face of those who would reserve the latter expression for the singular horror of Nazi concentration camps. This is a choice that shows evidence of having gone through the undecidable. Perhaps it was the sense that acquiescence in the face of those who seek exclusive ownership of this word would privilege one event of silent horror even as another one continues, at the dead end of other country roads, unacknowledged, largely unsung, protected by new alibis, new myopias.

Returning more explicitly to global warming (or whatever we call it), the significance of language in grasping or hiding its significance is hard to exaggerate. So, too, are the opportunities for deconstructive engagement with language—both everyday language, the common discourses of legitimation, and the language of philosophy itself. While the percentage of atmospheric CO_2 rises inexorably, the narratives we construct to justify the things we do that contribute to this rise have continuing legitimacy. (George H. W. Bush: "The *American way of life* is not up for negotiation. Period." [1992; emphasis added].) The discourse in which employment opportunities trump sustainability has a certain independence from the real, especially where that real is fabricated in part from what is still over the horizon. Language lags behind the real and often distorts it, even if there are no perfectly proper words. A deconstructive disposition does not see language as a surface phenomenon we can set aside, but as a deep and fundamental part of the problem. It offers many ways of performatively challenging and displacing the language in which key aspects of global warming are often articulated.

Aporetic Temporality

Following up our discussion of temporal phronesis (Chapter 3), I claim that our ability to "think ecologically" rests on confronting the aporiae of time. Climate change, its tipping points, its "locked-in" temperature rises, the need to have acted yesterday—all present unprecedented challenges to our thinking about past and future, and what we might do "now." Especially if it is already "too late." How does a deconstructive disposition help?

For phenomenology, it was important to step back from objective worldly time to the internal time-consciousness that makes it possible constitutively. For Derrida, this is a misguided attempt to return to a subjective self-presence riven by linguistic and temporal difference in a way that undermines the very presence it seeks. Deconstruction, on the one hand, abandons any attempt at a post-metaphysical *theory* of time, but, on the other hand, it proliferates a slew of aporetic temporalities we would do well to take seriously.[16] There is an ongoing resistance to any seamless linear time of progress. Derrida replaces this with an im-possible messianic time of hope in which, as with a democracy-to-come, any literal sense of time seems to be converted into a certain (im-)possibility and openness, much as happens in Heidegger.[17] If Nietzsche displaces eschatological time with an eternal return that forcibly extinguishes any residues of cosmic teleology, deconstruction is happier with multiple temporalities not subordinated to any sovereign time (what Hofstadter once called strange loops), such as a past that was never present or a hauntology that can never fully repay its debt to the past and is always haunted by what it imagines it could forget. Derrida draws upon such an idea specifically in thinking through just what kind of Marx we could still inherit, a Marx who, though a materialist, still has room for specters. He treats Marx in much the way that Nietzsche urges us to treat history, as enabling critically received possibilities for the creative furthering of life, even as elsewhere he will contest the very opposition between life and death.

9/11 perhaps offers the most striking example of a cluster of competing nonlinear temporalities. It was said that it could not have been anticipated, and yet many did anticipate it. While it all happened on that one day on September 11, 2001, images of it were relentlessly repeated, all over the world, in the days that followed. And the meaning of this event took a while to sink in. Early Derrida had written that "the future can only be anticipated in the form of absolute danger."[18] It is hard to see how this can be generally true, but it fits rather well the devastating effect of 9/11. As we said earlier, if that could happen, what now? 9/11 fractured our sense of a benignly unfolding future. It was an event in time that ruptured the time frame in which it appears. Time becomes irreversibly complex, and we cannot think it without inhabiting such complexity.

At the same time as we become aware of the future as a potential site of danger and dramatic disruption, it also becomes clear that this is nothing other than the past catching up with us. Our past practices, none intrinsically evil or catastrophic, accumulate like DDT in eagles until they overflow into dramatic change. What comes at us, seemingly from the outside, such as the melting of glaciers, is an indirect product of our own agency. We need not lean too heavily on 9/11 as an example of a temporally complex event. There are plenty more directly significant *environmental* events—including the melting of the ice caps. In America alone, we have hosted the Deep Horizon oil spill (2010), Hurricane Katrina (2005), the Exxon Valdez oil spill (1989), the Three Mile Island nuclear meltdown (1979), and the Dust Bowl (1930s), to name only a few. In each case, the unexpected, the aftermath, the lessons learned and not.

Again, some ask whether global warming will really happen. Others reply that the future is already here; we are witnessing it without being sure what it is. The eyes of the crocodile have just broken the surface, and the crocodile deniers spring up everywhere. For with all this talk of the unpredictable, some of the future has happened, and it is well known that the critical 2 percent rise in average surface temperature, after which real unpredictability is predicted, is already "in the pipeline." Derrida's "always already" is not just a quasi-transcendental mantra. We cannot return to simpler phenomenological times (as with Husserl) in which retention, current awareness, and protention would be happily interwoven with thematic memory and expectation to shape the time we experience. Some of the ingredients may be the same, but baking them together need not result in a digestible dish. The real objection to the thought that aporetic time is the new normal is that it was already the old normal—we just did not realize it. If so, we can no more return to a simpler way of inhabiting time than we can return to simpler times, or a lost origin. We can at best acknowledge the shape of such desire (for "presence"), guard against its seductions, and invent new shapes of dwelling.

It is often said that we should adopt a "precautionary" principle with respect to future environmental damage. In the absence of complete proof, we should still try to prevent harm by acting on the best evidence. In a sense, this is obvious because we cannot have certainty about a future that has not happened and will only ever happen once. In all sorts of areas, we already act like this, taking out insurance, for example, as a precaution. It is something of a complement to being open to whatever comes. And it captures the spirit of a certain middle voice, or perhaps a double voice, blending or mediating between agency and receptivity. Into this mix, we need to inject the disturbing thought that while we may know what kind of earth *we* would want to inherit, our descendants may not actually conform to our expectations. Our great-grandchildren may not value hiking in the mountains. This raises the question of what it is we should be trying to preserve. But it also raises the question of whether we might cease to care. Both the "precautionary principle" and a commitment to sustainability presuppose our ability and desire to identify with beings somewhat like us, whether it be narrowly (white, middle-class humans) or broadly (complex life-forms) conceived. What if, like the Atlantic Conveyor, this stream of projective identification ceased to flow? This sounds unlikely, but if one asked again Levinas's question, "Are we duped by morality?" and, with Hobbes, came to see most other humans as irredeemably violent, shortsighted, and self-interested, would we automatically want to encourage this species? What if, beyond video games, we develop simple, nontoxic ways of directly stimulating the brain's pleasure centers, and everything we now know as culture, indirect means to the same end, is set aside as old school? Would such beings be worth saving? These issues arise within the framework of time in that they affect the way we inhabit our temporal horizons—for example, our capacity for future projection, planning, and hope.[19] They are not issues exclusive to deconstruction. But they expand and extend its sense of the Subject not as some transcendental constituter but as colonized by such desires for control, domination, and stimulation, and open to further colonization.

If the future is a site of profound anxiety, the past is not entirely different. Those who resist the theory of evolution often do so because they cannot stomach the thought that our ancestors were monkeys.[20] And while there are some who would remind us that we are made of stardust, there are others (like me) who think of that as bad (reductive) materialism. In what sense is the history of the earth "our" history?[21] Does this extend through to the emergence of primates from mammals (sixty-five million years ago), or early versions of man in Africa[22] (some 2.8 million years ago), or only as far back as Homo sapiens (250,000–400,000 years ago)? Or do we see ourselves as part of the stream of life, itself developing from mineral existence, through organic molecules, to the earliest life-forms? Is the Big Bang part of our history? It is not difficult to see what gentle pressures have elicited such questions. We used to be located, roughly speaking, within geological history. Now we have hatched the idea of the Anthropocene, which would mark the unprecedented impact of a life-form on the geological forces of the planet (if we exclude the first oxygenating cyanobacteria), a confluence of two quite distinct temporal scales and streams. And we need some such schematization of multiple semi-independent temporalities to think through what we might call the uneven development of humanity. We have productive and destructive powers that outrun both our brain development and the political institutions needed to manage them.[23] Humans are notoriously bad, for example, at risk assessment when thinking about the medium to long-term future, consistently underestimating the risk of high-cost, low-likelihood events. There may be good evolutionary reasons for this. Ancient man had a much shorter life span. International law and bodies such as the United Nations are obvious resources both for preventing conflict and for combatting climate change. Yet their power still rests on the support, or at least the acquiescence, of individual states. And they are still prey to all the paradoxes of collective action they were designed to overcome. We need better brains and better mechanisms of collective agency to cope with the powers we have developed. Time, as Derrida said, quoting Hamlet, is out of joint.[24]

We Animals

If we (humans) can still use the word, we are and are not animals. ("What a piece of work is a man, how noble in reason, how infinite in faculties . . . the paragon of animals . . ." [Hamlet].) Biologically, we are animals, and yet "animal" is the name we use for a vast array of nonhuman fellow travelers on the planet. To call an individual human an animal is usually to denigrate or prepare him (or her) for inhuman treatment. To call an animal an animal is often to prepare it for dinner. Deconstruction takes as its starting point the ways in which our self-understanding as humans rests on the construction of the animal as a subordinated other. To the extent that our grasp and treatment of nonhumans is the site of the ongoing and probably interminable operation of the anthropological machine, both our experience of nonhumans and the ways in which we try to think about them require constant vigilance, lest we merely use nonhumans as projective screens.[25]

I argued some time ago that there is no such thing as an animal.[26] There are aardvarks, antelopes, armadillos, Australians, and there are vast differences between them. But is there not an abyss between man and animal, as Heidegger insists? It would be crazy[27] to deny a gulf, but it, too, is not one thing but many and varied. What all this argues against is any uniquely privileged hierarchical table of species, any attempt to covertly attribute general normative rankings to certain key characteristics. It demands a step back, however difficult that might be from our understandable tendency to value what we humans think we are good at. To speak of the human subject in terms of carnophallogocentrism[28] is to begin to constitute this being in terms of deep desiring practices, including meat-eating. There are obvious dangers in any reductionism, and yet the rewards—including glimpses of new shapes of thought—are quite real. In this vein, one might speculatively propose a new natural history of philosophy—as a rationalization of power first over other humans, and then specifically as a justification for kinnibalism.[29] Derrida insists that vegetarians do not escape the carnivore label. We "eat" others in countless other violent ways, and refraining from meat can be an alibi for a broader blindness to violence.[30] Deconstruction *need* not make any such specific claim, but it can track (and interrupt) the performative ways in which our use of language and its underlying schematizations sustain such claims.

There may be no one "question of the animal." We mentioned before how Derrida's ruminations on his cat in the bathroom, allowing himself to be put in question by its gaze, have implications for "humans" and "animals" in general (e.g., about who is more naked).[31] But he is insistent that he is speaking of this specific cat (while strangely supplying no name). The ethical may indeed be born from such singular encounters. And yet, as we have seen, and in his earlier arguments for a hyperbolic responsibility for all cats even as he feeds his own, Derrida does not hesitate to cast his net more widely, eventually addressing animal genocide and even the ways in which it exceeds the holocaust by breeding animals in order to be killed. Taking all this seriously would give dramatic new life to Heidegger's discussion of being-toward-death. It is important, too, to take seriously the way in which deconstruction puts pressure on the distinction between active and passive, the temptation to treat self-conscious agency as the paradigm of responsibility. For the *other* animal genocide, the sixth extinction,[32] is not the result of a conspiracy of evil but of creeping negligence. First, we didn't really know what we were doing, then we didn't want to know. To continue in the same way—with habitat destruction, with threats to countless species from climate disruption—is culpable negligence. In other words, while deconstruction can contribute to our thinking of animal rights by liberating the gaze of the individual animal, it also lubricates the path to environmental ethics by preventing us from hiding under the bush of individual agential responsibility.[33] We, indeed, are responsible. But it immediately raises the question of who "we" are. And this is not (if it ever could be) "just" a semantic issue. The "we" question has itself to do with how any such collectivity is constituted, by whom, to what end, and with what powers. And the "we" that might be needed to effect a change in "our" treatment of those "we" call animals, whether directly or indirectly, would need to be set aside if "we" came to include nonhumans in a broader we—this time a "we" of interdependence and common fate.

The interdependence of humans and nonhumans is often invisible. Few study the role of beetles or fungi in recycling dead trees and leaves. And yet without them, the cycle would cease, and life on Earth as we know it would grind to a halt. Might we not come to understand Derrida's democracy-to-come as (impossibly) embracing "animals," perhaps in alliance with Bruno Latour's parliament of all beings?

There is an invasion of the "we" on the horizon; both ethically and ecologically, we humans are not as separable from other creatures as we would like to think. Traditionally, we have tended to suppose that when it comes to microorganisms (e.g., bacteria, fungi, viruses), we do need to draw the line just to survive. Disease-causing organisms are simply the enemy. But again, deconstruction is well positioned to articulate the difficulty of this position. It looks increasingly as if an oppositional stance is a miscalculation (as in so many other areas). Consider the impact of antibiotics on human health in recent decades. Overprescription, failure to complete the course, and the routine use of antibiotics in meat production (faster weight gain) have bred drug-resistant bacteria—a phenomenon allied to that of the autoimmune response in its staging of the collapse of oppositional logic: two different failures of sovereign control. Moreover, the microbiome project makes it clear that what I call "*my* body," the one that must protect itself against the alien invader, is always already a we, crammed full of benign bacteria and other microorganisms, without which I could not, for example, digest. Broad-spectrum antibiotics bomb wedding parties while being aimed at terrorists. We do not know what we are doing to ourselves because we do not know quite how and how much we are a we (or many we's).

We selected three major deconstructive motifs to at least begin to show why the environmental crisis deserves to be treated as the eleventh plague. Sovereignty, topological complexity, the New International, inheritance, and death could be given similar treatments. The question we asked earlier was whether this could just be added on to Derrida's original ten, as if it had been overlooked or forgotten. Or is it perhaps the root of all the others? The answer to this question is important.

First, we need to remind ourselves that his ten plagues (*Specters of Marx*) are plagues of the New World Order, the one triumphantly celebrated by Fukuyama. In other words, they represent the dark underbelly of the free enterprise, free market world that, albeit imperfectly, is said to have brought so much prosperity to so many. Some of the headings are economic (unemployment, economic war, the burden of foreign debt, contradictions of the free market) but not all. There are various other dimensions of the military-industrial complex (nuclear weapons, the arms industry, and interethnic wars), and then there are failures of democracy (phantom states such as the mafia, the exclusion of refugees from the democratic process, and the broken promise of international law and institutions). Let us admit that there is much that is left out (e.g., persistent poverty, growing inequality, the power of multinationals)—this list is neither complete nor homogeneous. But how would global warming fit in? Can it just be added to such a short list of neglected plagues?

One reason to resist such an additive approach is that there are intimate connections between some of these first ten plagues and the looming climate catastrophe. Joining a disconnected list would be a missed opportunity to pursue those connections. For example, employment issues are some of the most urgent governments have to deal with and are specifically used as reasons not to pass or enforce environmental legislation. Intense global economic competition accentuates the tendency to externalize every possible cost—seeking states and countries with lax waste-dumping laws, precipitating a rush to the bottom. Nature (in the shape of rivers, oceans, and the atmosphere) then picks up the tab. Interethnic war is often fought over scarce natural resources, feeding the insatiable monster of development—the spreading demand for a better lifestyle. This then generates refugees, fleeing from war and destruction. Global warming will accelerate these displacements as people abandon desertified land, with all the tragedy and squalor of life in camps with poor facilities for those who survive. Foreign debt is a crushing burden on a country that deprives it of the economic surpluses that would enable investment in alternative energy sources and encourages harmful short-term choices. More generally, poor and deprived people have a greater interest in surviving until tomorrow than in embracing sustainable lifestyles. These interconnections and more suggest that merely adding number eleven to the first ten is not the right answer. However, there are two further levels at which we can pursue this question.

First, it is eminently plausible to think that without the kind of transformation—revolution—in the shape of human desires, and hence lifestyles, we are either doomed environmentally or face a return to feudalism or military subjugation along North Korean lines. These latter options would enforce poverty for the masses while the

1 percent live high on the hog, which would cut the average energy footprint. If we accept that the alternative is either doom or dreadful social and political regression, the prospect of real social justice, the realization of so much unfulfilled promise would cease to be some sort of felicity and become a survival necessity. We could begin to envisage a convergence between social (and interspecies) justice and environmental necessity. Much of what deconstruction has done already in terms of welcoming the questioning and revisioning of borders, boundaries, limits, and identities, and exposing the costs of modes of thinking, speaking, and dwelling that hide the costs of constitutive exclusion, would facilitate such a convergence.

Derrida's remarks about the animal holocaust, and about human suffering and misery, are set in the context of our denial, blindness, and refusal to acknowledge these phenomena, and the way that human suffering especially represents the contradiction, the hidden waste, produced by an ever more efficiently functioning system.

There is, however, a second step to be taken, which would make it even clearer that the eleventh plague is not just one more plague but something of a supplement that completes what seemed already to be complete. The naive and unguarded way of putting this would be to say that climate catastrophe threatens the material ground of our being and so cannot be compared to the other ten plagues, except perhaps the much more uncertain prospect of nuclear war. But does deconstruction have anything special to say about this, or are we stretching to make a connection? As I see it, climate catastrophe would be the "material" face of the culmination of an emergent contradiction captured in such key deconstructive concepts as the exclusion of the Other and the autoimmune response. The first tells us that establishing and maintaining binary dominance requires the suppression (ideally disguised) of the lesser force, an operation that will have unintended consequences. The second tells us that attempts at protecting a rigid identity will tend to destroy the very identity they seem to serve. Obviously, these are closely connected. To these two we might add a third, from "outside" deconstruction— that profits, economic value, on which so much of the world turns,

rest not just on the exploitation of wage labor but on the extraction of finite natural capital and the externalization of the costs of
"waste" disposal onto "nature"—especially such sinks as sea and
air. These three steps comprise a kind of logic. At this point, warning bells go off. The detachment of just such a logic from material
historical social conditions was Marx's objection to Hegel: *Geist* is
properly understood through "relations of production." Are we suc-
cumbing again to a kind of idealism, with deconstruction supplying
a quasi-transcendental underpinning to historical inevitability? Isn't
the lesson of such prognostications that we discount what we cannot anticipate? Capitalism did not wither under its contradictions
nor did it bring about the progressive immiseration of the poor. It
bought off its contradictions, at least for a while, and fanned mass
consumption into a force for economic growth. And communism
did not lead to the withering away of the state.

So, to clarify, this third step, this attempt to show that global
catastrophe is not just one plague among others, is not an exercise
in counterprovidential history nor an attempt to draft deconstruction into the business of prognostication. Rather, it marks the site
at which, it would seem, some of the fundamental bioexistential
parameters for human and nonhuman flourishing might well be
breached, and for all intents and purposes, irreversibly. It does so
by highlighting the aporetic realities of our thinking and dwelling.
Can deconstruction, however usefully, contribute to our thinking
about hurricanes, the melting of the ice caps, the release of vast
quantities of methane from Siberian permafrost, new disease vec-
tors, mass migration, starvation, and agricultural disruption? And
even if it can, do we really need it? Derrida was perhaps more comfortable reminding us that 9/11 terrorists attacked, inter alia, the
global communication network that made even their own terrorism
possible by broadcasting those images, metastasizing the terror. He
is perhaps less comfortable thinking of the earth as a whole, even as
a complex system of differences, with all the dangers of totalization
that that entails. For all his hesitations about the language of rights,
and its dependence on traditional notions of the subject, agency, and

responsibility, he does in the end strongly defend that discourse. I cannot think that he would treat with deconstructive aloofness the prospect that human flourishing might in the course of time be replaced by a life-form with an impoverished trajectory. I am left, then, with the question of whether deconstruction is of any direct help in adumbrating a kind of "(quasi-)transcendental" materialism, a materialism that would "ground" what matters as its condition of possibility, even as it occupied the same surface on the Möbius strip. Do we need to form an alliance with what has come to be called New Materialism?

Exploring these three threads only gestures at presenting deconstruction as an eco-friendly disposition. A fuller account would consider hospitality in the face of mass migration, welcoming the Other, the New International, expanding the idea of a democracy-to-come to include nonhumans, the ineliminability of enlightenment values, thinking the impossible, the paradoxes of both autonomy and collective action, and shared sovereignty. Each of these so-called topics is, in fact, the name for a dispositional exhortation. Wittgenstein once wrote, "[T]he meaning of a word is its use."[34] And Heidegger, "The point is not to listen to a series of propositions, but to follow the movement of showing."[35] Derrida is indeed what Rorty called an edifying philosopher, recommending—in his case—patience, resisting schematizing formulations, noticing the silent ways in which old binaries frame problems, taking the road less traveled, releasing the power of the repressed Other. Do we need deconstruction? Many a philosophical position can be characterized as an approach, or a way of thinking. Even a disposition. Climate change is a real physical phenomenon, the focus of much scientific research. But the human (cultural, social, political, economic) factors that lie at the heart of its seemingly inexorable path are not only the proper object of strictly philosophical reflection but call for the most thoughtful deployment of all our broader practical and theoretical resources. Deconstruction is not a method, not an algorithm, not a recipe, not a formula but a complex disposition—a resource we need when addressing the eleventh plague: anthropogenic climate change.

PART

II

Experiential Pathways

5

Things at the Edge of the World

A snake came to my water-trough
On a hot, hot day, and I in pyjamas for the heat . . .
And yet those voices:
If you were not afraid, you would kill him!
And truly I was afraid, I was most afraid,
But even so, honoured still more
That he should seek my hospitality
From out the dark door of the secret earth.
                    —*D. H. Lawrence, "The Snake"*

Confronted by the snake, an emissary of the strange, D. H. Lawrence is conflicted from the beginning, switching in a trice from fear and hostility to wonder and hospitality. Eventually, he throws a log at the snake: "And immediately I regretted it. I thought how paltry, how vulgar, what a mean act! I despised myself and the voices of my accursed human education."

This structure of switching or reversal appears in many places. It is found in Levinas's account of coming up against the limits of my own intentional orientation, its interruption by the face (or appeal, or call) of the Other, itself putting a strange reverse spin on Sartre's account of the effect of the look. It is found in Rilke's description of the experience of being looked at by a tree, in the various meditations (from Plato, to Hegel, to Nietzsche and Bataille) on the significance of the sun and in philosophical struggles over the place and status of the body—confined at different times to the position of sheer matter, burden, or instrument, and also at times released from this bondage with the recognition of its power to constitute the real. It is found in Heidegger's world-revealing meditations on "The Thing" and in his and Gadamer's reflections on the work of art.[1] In each case, a thing that begins as an object of experience becomes the site of an event of reversal and transformation in which not only the subject is implicated in an unexpected way but the world, or a part of it, is poised for restructuration and for the proliferation of new chains of possibility.[2]

I explore here the intriguing possibility that the world (as we call it) may be populated with beings of various sorts that in all sorts of different ways open worlds, open onto worlds, and open our eyes to possible worlds by interrupting this one. When Alice ventured into the rabbit hole, she discovered a world within a world. I am proposing that the world be viewed as a veritable rabbit warren, in which the entrances to these other worlds are marked by what we call things.[3] The world is no longer a collection of things in the ordinary sense, however heterogeneous. Rather, it is a space that enables spaces, a time proliferating times. And things come to be seen as events, sites for transformation.[4]

To rest contemplatively on a thing is to open up the world or worlds into which it invites us. Conceived in this way, things are marked by analogues of what physicists call event horizons. This is the point or line at which we switch from seeing the thing as in the world to seeing the thing as projecting, opening, or proliferating its own world, its own order of things, or as constitutively implicated in the world in which it might seem just to be an item.

I offer here an introduction to this project, locating within this account both human and animal strangers, and indeed recurrent productive strangeness. Part of my purpose here is to argue that the ethical dimension of such reversals needs to be set within the broader context of our flickering in and out of a whole panoply of strangeness. In the next chapter, I extend these considerations to the earth itself.

Let me begin with an example of such a reversal—our experience of the sun. It is a remarkable fact that this source of so much light should be dangerous to direct sight. We may quite properly understand this danger physiologically—that we could burn the retina. But there is another danger of a quite different order—that we may "see" something new, and disturbing, something other than what was blindingly obvious. The sun as a danger to those who live in the shadows is at the heart of Plato's allegory of the cave. But a more material revelation awaits us: recognizing the sun as the source of all energy, indeed of our sustained existence. For those of us wedded to autonomy, this may be an unwelcome reminder of dependency. And for those who have come to embrace a transcendent deity, the tantalizingly ambivalent status of the sun, as both an item of furniture in the heavens and the energetic ground of our being, might well cause a ripple of unease, at least in the decision to abandon sun worship. Maybe, for all their barbaric practices, the Aztecs were on to something. Might not sun worship be a practice of meditation on the profound significance of the sun and all things solar? And there is more to come. When we look at the sun, it is true—the sun does not look back. This gives us a certain assurance of privilege. And yet our eyes—what are they? They are the evolutionary product of living on an earth flooded with light for millions of years. These things I call my eyes spring from the sun, as does every animate organism. What looks at the sun is a child of sunlight. This is true of me as a living being and of my eyes as essentially attuned to the sun and to the visibility it opens up.

It is easy for phenomenologists to think of the "constitutive" in a formal or transcendental sense. But here it is importantly material and historical. The event of reversal or fracture is the one in which the I/eye that sees the God/sun comes to grasp it further as the condition of its own capacity to see at all, indeed to be at all.[5] We may confidently surmise that this is a seminal event, one that ushers in further developments. Such a dependency has what I call terrexistential implications. What would it be like to welcome such a revelation? Or to refuse it? What would it be, as Nietzsche asked with regard to eternal return, to affirm it, to will it?

Consider for a moment one of the deepest and most difficult aspects of this dependency—our being tied up in an unthinkably deep past. The time of evolution—of life, of animal life, of human life—is unthinkably deep both because it exceeds our capacity to imagine even in terms of scale, raising really profound questions about what those limits are, and what we mean by "imagine." Can we "imagine" the Big Bang? Or the end of the world? And beyond the question of magnitude, we must also imagine our not being here and indeed there being nothing like us on the planet. We have to peer back an interminably long way around evolutionary corners that block illumination from the distant past.

Exploring the temporal aspects of our dependency on the sun has just begun. A meditation on this opening soon realizes that our entire fossil fuel economy and way of life consists of releasing at an accelerating rate the reserves of solar energy locked up in gas, coal, and oil, burning it like there's no tomorrow.[6] We are tapping our stored solar past, as well as drinking in today's light and heat. Moreover, one day the sun will explode and the experiment will come to an end.[7]

The second example of reversal is the nonhuman animal. I opened with D. H. Lawrence's poem "The Snake," in which he laments his reactive violence. You could also think of Aldo Leopold's account of the light in the dying wolf's eyes.[8] Or Theodore Roethke's basking lizard, to whom he says the stone terrace belongs.[9] The movement, the reversal, the transformation in each case is one in which the animal moves from being a part of my world, our world, to making a claim, to occasioning on my part the recognition that the animal, too, in some sense, has a world and projects significant dwelling.[10] Some such experience takes place when I pick up a worm crossing the path after a rainstorm and deposit it safely on the other side. I recognize both vulnerability and something of a manner of life, or way of dwelling on the part of the worm. In modest fashion, I go out of my way, as we say, to help him on his way. Something happens here, but it is mostly an accommodation of one mode of life to another. There is no threat from the worm, and he never looks back.

The reversal has not yet happened; we are still on the cusp. Reversal here has many portals—we notice the spider's web, or the beaver's dam, or, as my ex-squirrel-hunting friend recalls, we watch squirrels playing on a tree trunk. In each case, a concept appears on the other side of the line (home, territory, play), and we are primed for something abrupt to happen. Another friend's advice on encountering wasps is apposite here: never make eye contact, and they won't sting you. It took me a while to realize this was a joke. But this possibility of eye contact, of seeing oneself being looked at, takes us closer to the edge. The reversal happens when we see ourselves being seen, and then realize that we don't really understand how they see us, and that we are as much a part of their world, whatever that is, as they are of ours. This reversal through an encounter with the other-than-human can come in ways that are all too sharp and clear; Sartre's account of the look has this abruptness about it. And being the object of the other's gaze is no mystery. In the movie *Grizzly Man*, Timothy Treadwell loses his life when he (and his girlfriend) are eaten by a rogue bear after thirteen seasons of peaceful coexistence in Alaska. The director, Werner Herzog, comments that Treadwell just did not get it: the bears saw him as food. And when he was eaten, he became part of their world in a very literal sense. This suggests not that bears have no "as such," as Heidegger might say, but that that their version, at the very least, even if it overlaps our own, may distribute value differently. But in some ways, this reversal is still too straightforward, like man bites dog, hunter hunted. Everything changes, and yet the conceptual space is still the same. Animal fables fall under this heading. More disconcerting, and almost completing the reversal as event, is our recognition that we may well have little or no handle on the other-than-human creature's "world." Heidegger himself seems to be of two minds in thinking this through. Disputing Rilke's valuation of the open, he insists on the animal's world lacking a certain disclosedness (tied up with truth), and he seems to go along with von Uexküll's references to functional tone when speaking of a "disinhibiting ring."[11] These accounts make the animal's world fairly transparent to us, as a reduced version of our own. But at other times, he suggests that we don't really know how the world seems to another creature.

Even that move, however, is but a stage on the way. The true reversal would come were this path to converge with the path of "What Is Metaphysics?," where Heidegger speaks of the "totality of what is" slipping away from us (in angst), the point at which we are reminded of the *unheimlich* at the heart of our dwelling. The animal, I am suggesting, has this power: to relieve us of our habituated dwelling by bringing us face-to-face with a significantly different and unassimilable mode of dwelling.[12]

The third case of reversal I will take is that of the other human, and here I divide the example itself into three: the sexual other, the stranger, and the enemy. The textual ghosts floating in the background here are Irigaray, Levinas (with Derrida), and Schmitt. Each of these relations marks the site of a radical transformation or is open to such a possibility. In pursuing these three cases, I will be able to differentiate my position from one centered on the ethical opening. Moreover, I believe it will be possible to explain the misunderstanding generating what we have come to think of as the ethical infinite.[13]

The sense of wonder is the mark of the philosopher.

Plato

Consider first the sexual other and Irigaray's insistence that the experience of wonder properly applies to the sexual other before it applies to the cosmos, thus upending any impersonal metaphysical primacy of wonder, such as Plato proposed. Irigaray in particular alludes to Descartes's account of wonder as the first passion, being moved by our first encounter with a thing before we know what it is. She asks that we return "this feeling of wonder, surprise, and astonishment in the face of the unknowable" to its proper place—"the realm of sexual difference"—by which, somewhat traditionally, she means the difference between man and woman. For Irigaray, wonder is the antidote to any claim to possess or control the other.[14]

How does this account fit into the schema I am proposing here? Wonder, or astonishment, is the experience that arrests and reverses, one might say, the everyday, possessive projective orientation to the sexual other, allowing us to treat and see the other as something of a miraculous complement, as if for the first time. Here it is not that the other looks back with the recrystallizing impact that Sartre describes in the gaze. Rather, a space of nonpossessive delight is opened up. Irigaray goes on to nominate angels as inaugurating and protecting sexual encounter as celebration beyond any master/slave drama of domination.

Da-sein is the happening of strangeness.

Martin Heidegger

For the second encounter with the human other, I take the example of the stranger, with an eye to Levinas's focus on the widow, the orphan, the stranger, and Derrida's meditations on hospitality. In the stranger, three different dimensions come together. First is the absence of knowledge: nothing is known about this person, whether he or she is to be trusted, well disposed, and so forth. Second, the stranger is in need, being away from home. And third, it is unlikely that you will meet this person again. The appearance of the stranger may occasion no response (drive on by, don't answer the door) or a negative one (closing the door in his or her face, turning down his or her visa application). Such responses would confirm a certain default self-centeredness. But there is also the possibility of the event in which I am taken out of myself and even moved to the point of what Derrida calls "pure hospitality"—a welcome without limit, and without checking credentials, in which I put myself (my house, my family, my country) fundamentally at risk. I see myself as hostage to the other without any consideration of reciprocity. The scales of native narcissism fall from my eyes, and I am exposed to the need of the other. I discover saintliness.

> Tell me who your enemy is and I will tell you who you are.
>
>   Carl Schmitt

For the third example, I take the enemy, in which the transformation works in the opposite direction. Sometimes, seemingly from nowhere, cooperative or at least tolerant relationships with my neighbor, my friend, my compatriot, or some person or group with whom I get on well break down utterly, to be replaced by the other acquiring the status "enemy," in extremis, to be shunned, injured, or killed. This happened in Bosnia, in Rwanda, and it happens in small ways on our streets and in late-night bars every day. It is the world of suspicion, of paranoia, in which any small event is scrutinized for its secretly hostile significance. It is a world in which Jews, communists, and gays are rounded up and killed. And a world in which one state invades another sovereign state in pursuit of narrow self-interest.

To this list—lover, stranger, enemy—we could add the friend, the master, the ghost, and many others. In each case, this category of the other can emerge or subside without reason, and by such transformations the world is transfigured.

The point of supplying these various examples, worked and unworked, is to demonstrate that the phenomenon of renewal and transformation in the self/other relationship is far from being restricted to the charismatically ethical cases of the other in need—widow, orphan, stranger. And the reason for this provides an ontological ground even for those privileged ethical cases, undermining Levinas's claim that ethics is first philosophy.[15] Let me now explore this thought.

I have described things at the edge of the world as sites at which events of reversal and transformation take place. And that the world opened up by this whole analysis is one of fractal space and time, one in which things turn out not merely to furnish "our" world but are open invitations to pass over into other worlds, rabbit holes. Focusing on the worlds these things open up takes us away from "subjectivity." And yet issues of that order are clearly at stake. These reversals all seem to involve a change in the direction of intentionality. Even if one wants to promote a fractal world, a space and time of discontinuous regions, one has to concede that access to such worlds is regulated by what might be called the fluency of selfhood.

I suggest that the entire domain marked by these events of reversal and transformation is generated by the combined operation of three different phenomena: (1) the primordial constitution of selfhood, (2) variable modes of identification with that self, and (3) the projection of modes of otherness consistent with one's manner of self-relatedness.

We could describe these events as involving counterprojection, perhaps a cousin of the phenomenological epoche, in which we cease our thetic possessiveness and allow ourselves to be guided creatively by what we might call play spaces. I am imagining here a generalization of Gadamer's sense of entering the space opened up by a work of art.[16]

But our analysis can take another direction. If the shapes taken by the self/other relation are such as to fortify a primordially constituted self, the question remains—from what "material," with what ingredients, is such selfhood constituted? I do not want to suggest anything very new here, except to say that philosophers and phenomenologists need more regularly to take on board that the fundamental matrix of self was constitutively relational from the beginning. By this, I mean that there can be no getting away from the original drama in which a human infant arrives on this earth utterly dependent on others to satisfy his absolute needs—especially hunger and sociality. The site of such needs is not yet a self, but the space in which such needs are met or frustrated is surely the original matrix of selfhood—laying down ground-level assumptions about whether and how quickly the world responds to my expressed needs, providing an original formatting to rhythms of desire and satiation, and expectations about my ability, through social interaction, to affect these outcomes. Does a breast appear when I call out for it, when I need it? For Freud, the anxiety of this experience of infant dependency feeds subsequent religious belief. In the twenty-first century, giving proper weight to these issues is no longer the special concern of psychoanalysis but of any theory that takes seriously the layeredness of human temporal constitution.[17]

As I see it, the manner and the upshot of these transformative reversals has to do with whether and how the original "material" of self-constitution, my prereflexive formative relationships with my early caregiver(s), is enabling or disabling.

Let me give an example: It appears that for decades, exposing the bodies of pregnant women was actually banned from American movies under the Hays Code.[18] It has been suggested that this is because of the implied sexuality thereby betrayed.[19] But it seems just as likely that it betrays an original anxiety—that we (men, especially) who are trained as autonomous beings find it difficult to acknowledge ontological dependency. And yet this dependency is fundamental; the question is whether and how we acknowledge it. Wittgenstein declares that man is an essentially dependent being, and then adds: "And that on which we depend we may call God." Well, we may, but we may not. Irigaray clearly sees that move as a displacement of a primary passion whose proper place is the sexual other. Melanie Klein, on the other hand, would see the religious as a displaced refusal to acknowledge the constitutive role of mother. And this refusal is understandable if we lack the means to adequately conceptualize the event or process of self-formation, if, that is, we insist that there must be a self there from the beginning to act on.

I am claiming that the domain of these topological transformations and reversals is itself made accessible/available to us to the extent that we have some sort of affective access to the grounds of our selfhood, one compatible with dependency or, more specifically, constitutive relationality. Moreover, not just compatible! For we are surely immersed in this "dependency" from the beginning, and this dependency includes engagements with other living beings.

What we think of as the autonomous self is surely a convenient construction.

At a certain level of generality, this account of reversibility clearly connects with the chiasmic sensibility that Merleau-Ponty develops in *The Visible and the Invisible*, which generalizes from the relationship between touching and being touched, between the sentient and the sensible—a broader sense of subject and world, or self/other, as mutually implicated, a condition he calls "flesh." He displaces the privilege of the eye in favor of the hand—touching and being touched. In my example, the eye itself is already "touched" in the sense of being materially conditioned by what it sees in the shape of the sun.[20] What is "reversed," ruptured, is the sense of the virgin birth, the uncaused cause, the autonomous subject. It is a delightful paradox that it may well be that creative autonomy is precisely something that has conditions—perhaps the good-enough mother![21]

In a way that will take us full circle, back to the sun and the animal, I end with another case of reversal, one on a par with the Copernican revolution and one with both individual and collective significance.

Throwing a banana skin out the car window, pumping effluent into the river, tossing out all the old magazines, even trying to forget one's old girlfriends or filing away random papers under "miscellaneous," all have one thing in common—they are acts of faith in the power of Away. It used to be a place from which things and people did not come back. For the English, the rot set in when the descendants of convicts we sent to Australia returned to beat us at cricket. But right across the board these days, Away is not playing ball. We throw things "away," but the landfills are full. We pump effluent into the river, and drinking water downstream is contaminated. What used to be a straight line has become a circle, indeed a Möbius strip. We still throw things away, but the "other" side of the strip is continuous with this side. This is a fundamental schematic reversal, a dramatic shift in the shape of world relationality. The nurturing, sustaining world can no longer guarantee its capacity to play that role. Our toxic activity is affecting the earth's capacity to nurture and sustain us. That this could be happening is a major threat to a certain understanding of freedom and agency, one that was perhaps always mistaken.

The transformative events that erupt with a meditation on the sun, or on the animal (and on the earth), would help flesh out an evolutionary story of our constitution as humans. Coupled with the developmental story I proposed, which could be opened by a meditation on Mother, or Breast, or Infant, we have the ingredients for a story of deep material constitutive relationality, one that would doubtless disturb the traditional poise of our autonomous agency.[22] The next question, one of ethical significance in the sense proposed by Heidegger in his *Letter on Humanism*, would be whether and how we could will—that is, performatively embody—a self-understanding in which we are essentially processual, embodied beneficiaries of the evolutionary adventure, in a way that preserves rather than tries to assimilate the strangeness.

This scenario of fractalterity would replace that of homogeneous space either in the sense of my world, laid out before me, or that of a single space in which all world-holders find their place. The world of fractalterity is one that cannot be properly represented but rather primes us to expect displacements, reversals, and transformations. It is in this world that strangers, gods, and monsters properly flourish, not crowded onto the same stage as at the end of a play but as various manners of world-opening, mobilizing, as I have said, the deepest resources by which selfhood is constituted.

Without for a moment putting in question the ethical significance of those events of self-displacement that Levinas highlights, my claim is that the capacity to break out of primary narcissism is not itself specifically ethical but ontological, and shared by events in which imagination, aesthetic adventure, and erotic delight are center stage, rather than any ethical engagement. The stranger may well be in need, demanding the bread from my mouth. But he may equally, as with the stranger of Plato's *Sophist* (quoted by Heidegger at the start of *Being and Time*), disconcert us, question us, announce the *unheimlich*, and call on me to set out with him on a dangerous or disturbing path: "For manifestly you have long been aware of what you mean when you use the expression 'being.' We, however, who used to think we understood it, have become perplexed."[23]

What I am proposing could be seen as radicalizing the move in *Being and Time* from a metaphysics of subjectivity to one based on constitutive world relationality. But the version of being-in-the-world I am adumbrating here is one of fractalterity, in which we are essentially exposed to manifold ways of world-making,[24] to alien avenues, to portals of possibility. Through wonder, perhaps also through horror and disgust, we may find our friend, lover, or nation—or indeed ourselves—to be the strangest thing.[25]

Heidegger proves a rich resource here, in particular, his reading of Sophocles's *Antigone*, the first chorus of which begins, "There is much that is strange, but there is nothing that surpasses man in strangeness."[26] Heidegger will merely translate these thoughts into his own language, naming death as that "strange and alien (*un-heimlich*) thing that banishes us once and for all from everything in which we are at home." Man is "always and essentially without issue in the face of death. His Da-sein is the happening of strangeness."[27,28] And death, indeed, is perhaps the thing most at the edge of the world.

Against Levinas, I continue to press the idea that the ethical version of expropriation via the demand of the other is not fundamental. I suggested earlier that I had doubts about infinite obligation, that it was perhaps a misunderstanding. Why do I say this? The structure of being invited, seduced, made demands of, challenged by something that exceeds any representation is not limited to ethical demand. Indeed, it captures quite well the lure of the thing, which in projecting a world defies representation in any other world in which it is merely furniture. One could say that in the humility required to acknowledge such an excess, there is an ethical opening. But it would apply to death, to a work of art, to God, or to the sun quite as much as to the human other in need. The idea of the infinitely demanding has a heroic cast to it.[29] Derrida will ask how he can justify feeding his cat and not all the other (hungry) cats in the world, and insist that if one thought one could calculate one's obligation, one would have reduced it to an algorithm, a rule, and in effect be guilty of pursuing a good conscience. And the infinitely demanding may mislead us into giving centrality to the ethical example.[30]

Let me try to tease apart this bundle of thoughts; a number of strands are woven tightly together. First, there is the idea that my obligation is unrepresentable; second, that this obligation arises through a radical (absolute) schematic reversal in which the other is now at the center of the universe; and third, that this relation to the other is asymmetrical in the sense that there can be no relation to or dealing with any demand I might be thought to make on the other (cf. pure hospitality). An incautious weaving together of these strands leads to a heroic misunderstanding. The reversal of perspective—decentering self-concern—is absolute in the sense that the Copernican revolution was an absolute. Being the center or rotating around a center are topological options between which there is no compromise, and if we take the idea that no representation is adequate to the other and feed it into this absolute reversal of perspective, we seem to get an absolute incalculable and unfulfillable obligation to the other. There are (at least) two problems with this. The first is that while the reversal of perfect narcissism may indeed be sainthood, such a schema begins with an implausible account of our fundamental condition. Levinas is providing an antidote to a metaphysical position (being, being-for-self, persisting in being) without first critiquing it. On my account, the infinite is the "reversal" of an exaggerated account of our original self-absorption, or primary narcissism.

Second, there is a deep temptation to understand the incalculable, if not exactly quantitatively, as at least able to relate to the calculable in comparative ways: "No response I make could ever be enough," as one might say, suggests we need to do more, even if that, too, "would not be enough." But an unrepresentable or incalculable demand does not require that conclusion. It means merely that no concept or number can ever adequately represent the obligation we suppose we have. That could be captured by much more modest formulas such as "doing one's best in the circumstances," "doing what seems right," "doing what felt appropriate"—none of which claims to calculate adequacy. To go further on this reading would be to evince a traumatized inability to bring any measure to a circumstance in which the investments of narcissism have been turned outward and locked onto the other.

This whole account I am giving is open to objection. Does not its stress on multiple, often incommensurable "worlds" encourage social isolation, political parochialism, and so forth? And surely there is at least a tension between this fractal space (and time) that would resist synthesis, and the idea that the life-support systems of the earth as a whole are in peril, and humanity with it. Does not this latter require the very unified space that our fractal vision forbids?

First, this position is not calculated to meet some extrinsic political agenda. But, in fact, the broad, shared recognition of a fractal universe is one that might be expected to increase tolerance by highlighting the resistance of "the world" to oversimplification. A fractal model of space does not mean that we cannot meet and talk. It would, however, be consistent with the thought that representation, political or otherwise, may not be as straightforward as we might think, or as reducible to formulae. We need each other to constitute a self.

As far as the tension with the need for a globally unified vision is concerned, a longer story is called for. I have suggested that however much we may always have known this locally, the recognition that at the planetary level there is no Away, that garbage returns, that the earth is a Möbius strip, is itself an event on a par with the Copernican revolution. Underlying the belief that we will need to protect and sustain the processes that protect and sustain us in particular and life in general is the thought that while we cannot actually produce an inventory of these processes, cannot fully represent them as a whole, we can nonetheless read the signs: species loss, dead oceans, rising CO_2, violent storms, and the list goes on. A fractal world is compatible with being alert to and acting on critical indicators, even if the mechanisms, the processes, the subsystems they reflect are often opaque to us.

This chapter is a trailer for a more developed project, one that articulates a heterogenous understanding of "things at the edge of the world." It fastens on those experiences of reversal, transformation, and estrangement in which "things" conceived of in a broad sense break out of the box of a focused intentionality and invite us (or challenge us) to different ways of worlding. Our capacity to respond to such events is not unconnected to the shape of our investment, individual and collective, in certain modes of selfhood. True autonomy recognizes its own constitutive relationality and delights in exploring the space of fractalterity. The cultivation of the *unheimlich* more generally is already of ethical significance and precedes the ethical opening generated by the face of the other in need. The capacity for a certain self-displacement, the openness to Copernican shifts, turns out to be a condition not just for earthly delight but also for our sustainable existence.

Might not the earth itself be a thing at the edge of the world? Like the sun, it may seem to be a rather special, very big complex object that we can photograph from the sky, that we can shape to our needs, that we can study and control. But that way of experiencing it reflects a habitual practical disposition vulnerable to dramatic reversal. I pursue this thought more deeply in the next chapter.

CHAPTER

6

Reversals and Transformations

Experience comes in two distinct stripes: those that confirm the metaphysical or conceptual armature that seems to sustain them and those that precisely challenge those frames and schemas. Heidegger, Derrida, and a number of other thinkers once resisted talking about experience, perhaps focusing too closely on confirming experiences. But these reservations are not compelling if one thinks about experiences that challenge. Moreover, the fact that Derrida has shepherded the return of experience is revealing.[1]

In the previous chapter, "Things at the Edge of the World," I described various experiences of reversal. I argued for what I call a fractal ontology, one in which what we think of as the furniture of the world either has the capacity to project a world of its own or can be seen to be constitutive in some important way of the world of which it initially appears merely to be a part.

I would now like to ask about the significance of such challenging experiences, especially in the form of reversal, transformation, displacement, and frame-shifting—how they are possible, what they suggest for the practice of philosophy, and how they bear on the shape of our earthly dwelling. For such experiences frequently drag into visibility background conditions that are normally in the shadows. I treat these experiences in a loosely phenomenological way, albeit with a deconstructive twist. This is meant to highlight what we might call aporetic experience, which when carefully described,

121

discussed, and interpreted challenges the premises of a first-order phenomenology—which is not a bad thing. I do not thematize reversal as such here in a way that would give center stage to Nietzsche, Merleau-Ponty, and Levinas. I leave that for another time.[2]

Why focus on experience? No doubt oppression, torture, discrimination, occupation, imprisonment, enslavement, deception, humiliation, and theft of land can all inspire critique and revolution. I want to focus here on experiences that are more broadly distributed, more widespread, and that provide the persistent fuel for philosophizing. They provide support for the thesis that experience is the unquenchable fuel for philosophical reflection. As far as frame-changing experiences are concerned, perhaps when Plato and Aristotle wrote of wonder as the source of philosophical inspiration they were just highlighting a prominent member of a broader category.

The word "experience" (and *Erfahrung*) came into its own in the modern era and is coupled with a certain privileging of epistemology.[3] If Kant rescued experience from its restriction to the senses, opening it up to reflection and judgment, it could be argued that, notwithstanding the destabilizing possibilities inherent in reflective judgment, its scope was still limited to a largely static individual self, and that it would have to wait for Hegel to be opened up further to the collective, the dynamic, and the historical. Experience is not something we have but something we undergo. To the extent that, even so, it is animated by a drive to a certain teleological fulfillment, it could (finally) be claimed to be a concept captured by a program. In his inaugural lecture, Foucault was at pains to insist that Hyppolite (his teacher) "never saw the Hegelian system as a reassuring universe; he saw in it the field in which philosophy took the ultimate risk."[4] Experience's capacity to expose us to the radically other would require another turn, one with an uncertain return.

Heidegger's impatience with the word "experience" is evident in many places. He identifies it with a certain ontological complacency in which once one has a rich enough concept of the person or self, one can dispense with the disruptive impact of the kind of exposure implied by the disclosedness of truth, in which man is placed, as it were, in the service of something he cannot master. Thus, in *The Origin of the Work of Art*, he writes, somewhat dismissively: "Everything is an experience. Yet perhaps experience is the element in which art dies."[5]

The early Derrida was not sympathetic to experience, either. It can also easily be a proxy for the philosophy of presence.[6] But in his later work, he became more methodologically relaxed. In *Force of Law*, he wrote of the undecidable as "the experience of that which, though heterogenous, foreign to the order of the calculable and the rule, is still obliged . . . to give itself up to the impossible decision, while taking account of law and rule."[7]

Aporias is full of references to the experience of "the non-passive endurance of the aporia," the "necessity of experience itself, the experience of the aporia . . . as endurance, or as passion, as interminable resistance," the "experience of the threshold," of the "impossible," and so on.[8] It is my sense that this return of experience is an invitation to its reconceptualization. Clearly, Derrida himself makes strides in this direction, especially in dissolving or softening the boundary between what we might call the metaphysical and the ethical. When I turn to considering these issues, and their implications for philosophical practice, I will want to say more.

But first I want to draw further upon Chapter 5 of this volume and its basic idea that we often find ourselves witness to a strange metamorphosis in our experience in which something we had taken to be "in" the world comes to be seen as more than that, either as projecting a world of its own or as actually helping to constitute the very world in which we thought it could merely be located. I gave various examples, including the sun, death, and the other sex. Let me rehearse the example of the sun. It is a common experience, though less so in an English autumn than a in Tennessee fall, to notice the sun in the sky. There it is—we gesture upward—along with clouds, the moon, stars, birds, and planes. But it does not take long to realize, indeed *remember*, and it may happen in an instant, that the sun is more than a disc in our visual field, more than a large hot object a long way away. It is the material ground of our existence as living beings, of terrestrial life itself, of the possibility of our having evolved at all, of the light that makes anything visible, including the sun itself, indeed, of the eyes with which we see it. Finally, we note (and Heidegger notes, too) that it is only through its rising and setting that we have time, perhaps sleep, dreams, and so on. This growing realization is not just the assembling of information but an absolute and radical shift of perspective. An item of sky furniture is revealed as foundation, as ground.

It may be thought that this is too good or too strong an example. Should we not focus more on how we could only have come to think of the sun in that crazy reductive, objectifying way after thousands of years of human history testifying to its deeper importance—including sun worship? These "reversals" could perhaps best be thought of as recollections, recovering what we already knew—what Plato called anamnesis. A similar story may be told about our confrontation with death, especially the possibility/certainty of one's own death, not to mention that of loved ones and whole species of other creatures. And of our coming to grips with the existence of other humans, without which nothing—no language, no culture, no value. In these examples, the reversal takes a double form, from foreground to background and from mere item to deeply significant phenomenon. It is worth saying here that this kind of reversal is quite different from the facile inversions of value that leave the space of their deployment intact.

First, I will discuss a broad range of experiences of reversal, reframing, disruption, and transformation with a view to drawing from them some sort of productive schematization. These examples extend and deepen the cases discussed in the previous chapter, and I will draw together their implications for a transformation in human dwelling. Moreover, the possibility of philosophy itself rests on these experiences, and many of them are common, everyday phenomena, whether or not they are taken up reflectively.

In the broadest terms, we can distinguish experiences of reversal that are arrived at by or through reflection and experiences intrinsically constituted by a reversal.

Worked Examples

First, the broken hammer. This is one of Heidegger's most famous examples. You are hammering away and the hammer breaks. Before it broke, you did not think about what a hammer is, how useful it is, how much you depend on it. Now all that becomes visible just as the hammer becomes a useless shadow of its former self. The being of the hammer—its usefulness—is not a property of the hammer, such as its material or weight. It has to do with its connectedness to other tools, projects, and so on. As is often said, if you have a hammer, everything looks like a nail. The broken hammer reveals this whole network of relevances, just as the car that will not start brings home to us our dependence on engineering, roads, the accessibility of remote locations, and how heavy our shopping can be. Its role in Heidegger's thinking in *Being and Time* is to highlight the way in which being a subject is to be always already outside ourselves in our worldly engagement and hence to displace the very idea of subject, inside/outside and self/world.

Second, falling in love. "I feel the earth move under my feet" (Carole King). Wittgenstein writes that "[t]he *world* of the *happy man* is a different one from that of the unhappy man."[9] All the more so for the world of the man (or woman) in love, at least as it is often reported. It is fundamentally a shared world, lit up anew with meaning and vitality, a world of infinite promise and possibility, a joyful world. The experience of falling in love, as it is put, seems akin to religious conversion as William James reports it: infinite happiness, being embraced by God, overflowing benevolence toward others, total loss of anxiety, and so on.[10] In each case, we may say, with an accompanying affective intensity, that there is a dramatic reconfiguration of the boundaries of self and other. What Heidegger says here about attunement (*Stimmung*) is worth repeating: such experiences disclose to us that we are always already attuned to the world in one way or another, exposing the shape of that attunement—the strength and character of connection, whether one is alone or sharing it with others. The same may be said for Heidegger's very different account of angst in *What Is Metaphysics?*, the experience of *das Nichts*, of the totality of what is, as it slips away. And in slipping away, our constitutive connectedness to it becomes visible as such.

Third, "Man speaks in that he responds to language" (Heidegger).[11] Consider reading a philosophical text. In *What is Called Thinking?*, Heidegger speaks of two alternative ways of reading: going counter and going to the other's encounter. In the first, one has an oppositional relation to the text and/or its author. In the second (and the transition from one to the other really can happen as a reversal), one goes, as Heidegger says, to the author's encounter (with the question of being). To this reading encounter, he insists, one must bring one's own deepest question, so that one has, as one might say, real skin in the game.

On this new model, everything changes. Instead of being a two-term relation, it is a three-term relation (or four). The text is no longer just a surface but the result of a struggle, a struggle one is being invited to join. And it is not just personal (within the writer). Rather, it is a struggle for being, for truth, for disclosedness, one that, if we are lucky, we share. In a different language, Derrida writes of a writer writing in a language whose logic he cannot completely master.[12] A deconstructive reading will trace that incomplete mastery, without itself claiming such mastery for itself. Or again, Heidegger writes of our truly speaking (or writing) only when we learn to listen to the voice of language. And so we may say, as Heidegger does of Kant, and Kant says of Plato, that we can understand an author better than they understand themselves, not because we are smarter that they, but because we can thematize what is for them structurally unthought, but delivered to us as an opening by their thinking. The key reversal here is captured by Heidegger speaking of disagreements between great thinkers as lovers' quarrels. Of course that opens up a further discussion about the significance of lovers' quarrels. And it may be that a further reversal or at least a check or caesura appears when one comes to think one has reached the limits of language, as when Wittgenstein describes that point as ethics,[13] and Adorno insists that, after Auschwitz, poetry is no longer possible.[14] Shifts from language as instrument, to language as reservoir, to language as empty are profound and unsettling. To possess logos may be precisely to recognize that we do not possess it. As writers this may appear in the form of a certain linguistic reticence or experimentation (crossing out, writing under erasure, inventing concepts), paradox—man is not what he is and is what he is not, excess (Monique Wittig's writing on the lesbian body[15]), playfulness (Hopkins' poetry) and so on.

Fourth, "Do not go gentle into that good night" (Dylan Thomas).[16] Death, or the thought of death, is the name of another space of reversal. And I confess I am tempted not to speak about it for fear of merely "using it" as an example. But it would then become the elephant in the room. So here goes.

We can begin, again with Heidegger, by alluding to being-toward-death, which he links to Dasein's coming to terms with its ownmost possibility. I would connect this to Hegel telling us that as finite beings we have to be determinate to prevent our life from evaporating into mere possibility. This sense of mortality is not morbid or dispiriting but, as Nietzsche insists with regard to the eternal return, is potentially galvanizing and bracing. Derrida will cast doubt on whether such a sense truly marks us off from animals—we surely have no clearer concept of death than they do. But I do not think this is a plausible line of thought. Our sense of mortality does not simply fall short of adequate conceptualization; it is anxiously (or perhaps joyfully) engaged with that inadequacy. Or it can be. This awareness of death is not our everyday condition. And when it comes over us, it does indeed change everything, one way or another. Another analogy: there is a large coral fossil etched on a broad rock outside my house that my geologist friend tells me is 450 million years old. This very information brought about a massive temporal displacement! At that time, this whole area was a vast, shallow sea. I felt at the same time very small and insignificant, and also infinitely special in being able even to try to think such a span of time.[17]

When I say death is a space of reversal, I mean that within its prospect various distinct experiences are possible, in which we undergo a shift in our cognitive, affective, existential frame. My general line of thought here is that such experiences make intuitively or reflectively available both our everyday predispositions and what is at stake in our worldly attunement. These various levels of insight are further enriched by literal near-death experiences, by the death of those dear to us, and by witnessing, or bearing witness to, the deaths of others, often strangers. Clearly, it makes a difference what beliefs about death we begin with. The prospect of heaven or hell, or the transmigration of souls, could be a game changer. And it will make a difference whether these deaths are natural or the products of malevolence. To be all too brief, the kinds of recrystallizing attitude transformation these make possible include: viscerally recognizing one's mortality (moving from unthinking acceptance or theoretical grasping to a life-changing rebirth), a renewed recognition of the reality of the lives of others (that *you* are mortal, and this matters to me), losing the center of one's world (bereaved partners often die soon afterward), incandescent outrage at the perpetrators of genocide, shame at belonging to the same species, guilt at having survived, unspeakable horror that it happened, determination that nothing like it will happen again on my watch, infinite sympathy for those who died, and genuine remorse on the part of perpetrators. In each of these responses, any repose that one had before goes out the window, and an opening is created for other, more lasting dispositions. I would add that there is a link here back to what we were saying about reinhabiting language and Wittgenstein's reference to silence at the limits of language. The intensified sense of the fact of one's own existence after one of these skirmishes with death may be inexpressible. One may be tempted to say, for example, "I exist." But, as Rilke asked, "Who if I cried out would hear me?"[18]

Fifth, empathy. Piaget describes the developmental shift in a normal child's empathic capacity registered by an experiment in which a child is asked to say what a doll on one side of a hill can see. Up to a certain age, the child describes what she herself can see. Eventually she learns to take up the viewpoint of the other—she can put herself in the doll's place. Indeed, we speak more broadly of walking in the other person's shoes as an opportunity for a revealing displacement of our egocentrism. This example shows us that reversals need not be accompanied by profound reflective realization, being dumbstruck. The work is being done by the observer—the developmental psychologist, the reader—who grasps the significance of such shifts. Levinas's sense of the repolarization and ethical transformation of the intersubjective field in the face-to-face relation is another example of such a displacement. This experience is, of course, common currency to philosophers, but it is no less significant for all that. How, we might ask, is it possible? When we think about it, it seems clear enough that there is no one such experience but more of a class of such experiences. And some are quite convoluted.

It happens that I am writing in the Atlanta airport only a few feet from a display cabinet in which is exhibited the giant foot of an elephant capped off with foam-padded tiger skin wrapped over the open end to make a seat. This is part of an exhibit of six such trophies, all confiscated from travelers returning from Africa as part of an anti-poaching campaign. Immediately, and without reflection, my response is to try to put myself in the shoes of the hunter who killed the elephant in order to chop it into trophy parts. I fail miserably. How could anyone do that? The hunter imagined, correctly, that there are plenty of other humans—consumers—who would gladly pay for such a specimen. On further reflection, I imagine being the elephant being chased, being shot, dying. It is said that elephants understand death—are that elephant and I on the same page in this fundamental way? Do we constitute a we? Then I imagine a hunter, dispossessed of his land, desperate to feed his starving children by any means. And then those who put the exhibition together, putting themselves in the place of viewers like me who would share the outrage. Finally, there is small part of me moved by the size of the elephant's foot, guiltily grateful to have come to appreciate that from a safe distance.

So we have a potpourri of imaginative displacementsfractal—fragmentary, contingent—in the direction of various others occupying the intersubjective field. If we stick to the more benign and generous proliferations, with, for example, Scheler, we discover not just empathy but sympathy, mimpathy, and propathy.[19]

It is, of course, a cliché to connect such experiences with the development of ethical life, though the logic of that connection is seriously disputed. And it is an exaggeration to say that our standing position is that of a narcissistic egocentrism we need to be snapped out of. This example shows us, moreover, that while the space of the empathic is one in which immediate feeling is seamlessly extended and developed through reflection, nothing would happen were experience not structured in such a way as to be open to these reversals and transformations.

To adequately understand the going-out-of-oneself reflected in empathy, we need, in fact, to provide a broader schematization. Levinas's account of the neo-Copernican shift wrought by the face of the Other is a true reversal, arguably to the point of a new pathology (I am hostage to the Other). Ethical expansionism, however, need not proceed in this way. Seeing myself (seeing "man") as "plain member and citizen [of the land community]" (Aldo Leopold)[20] is to set aside a relation of opposition in favor of one of co-belonging. The movement Hegel plots (*Phenomenology of Spirit*) from unreflective individual awareness to the recognition, in the shape of absolute knowledge, that subject and object, self and universe, are one is an analogous structural shift.

Ethical expansionism seems to take a gentler path. If one begins feeling sympathy with all those in pain, one might be thought to be merely expanding the reach of one's sympathies to "extend" this concern to include strangers and, after Bentham, nonhumans. Expansionism and reversal differ in that the first seems like a process and the second an event. But if the process of expansion progresses, assimilating the "other," it will eradicate the basis for any abyssal difference that would need special bridging or an abrupt structural transformation to effect.

Sixth, the animal. "We reached the old wolf in time to watch a fierce green fire dying in her eyes" (Aldo Leopold).[21] I have a friend who recounts the last day he went squirrel hunting. He was fifteen and his father had taught him to shoot. He went into the woods, found his squirrel, lined it up in his sights, and was about to fire when a second squirrel showed up. You might suppose that he did not then know which squirrel to shoot first. But this was not his problem. He watched the squirrels chase each other around the tree for twenty minutes, then lowered his gun. Permanently. What happened? He came to see squirrels as social beings, capable of play and enjoying relationships. *That* he could not shoot. Or again, consider D. H. Lawrence's account of an encounter with a snake at a water trough. He throws a log at it, and the snake slithers away. He then feels disgust with himself and laments missing the chance to meet with one of the "lords of life."[22] He begins with a clichéd hostile response to a feared or lesser being, then reflects on his own folly and transfigures the status of the snake. Or again, Derrida's report of his being looked at, naked, by his cat in the bathroom. He realizes he cannot appropriate or domesticate the gaze of the cat.[23] Or again, Aldo Leopold's description of the fading light in the eyes of the wolf he has just shot. In each case, the experience consists in the overturning, the reversal, of the gaze that would subordinate the animal as occupying a lower rung on the great chain of being.[24] These reversals, and I believe this is true more generally of my examples, are not mere inversions that would reinstate the frame of reference in which they occur.[25] Rather, they more closely resemble the operation with which Derrida originally introduced deconstruction, in which a reversal is accompanied by the introduction of an *indécidable*, a word or at least a term that cannot be accommodated within the frame at all.

To these encounters with animals, we can add the textual experience of watching the very word "animal" skulk at the corner of the page. We can see, or come to notice, that it is not just a name but a license to kill, or at least subordinate. And *that* thought sets up reverberations through language—how much more generally true is this with other expressions? Perhaps Derrida is onto something with his *"animot"*![26] And then again, consider the dehiscent experience of looking at one's own hand, or pulling out and inspecting some of one's own hair, noting one's own essential animality, even as one's own noting, as well as the symbolic privilege of the hand, makes that word tremble.

Seventh, and penultimately, if irresponsibly, I must say something about art. Irresponsibly because one cannot begin to say enough. Indeed, the literal experience of art might be exemplary of what I have been talking about. Does not art worth the name disclose, reframe, transform, and throw us back upon ourselves? Do not beauty and the sublime both delight us with their objects and yet teach us about our own capacities for shaping and bringing together? But let me for a moment speak of art in the terms I used in Chapter 5. I suggested that some of the things we encounter in the world are not so much *in* the world (though a painting may be hung on the wall, a sculpture may lie in the garden, a dance may be performed in the courtyard, and a novel may be shelved in the library). They also open worlds of their own, or special aspects of this world, as if they were variants of Alice's rabbit hole. What that means is that they have a certain autonomy of signification, into which they draw you, or which they project onto the world. A landscape painting may project the promise and prospect of a conquering people over a new land (Thomas Cole), a sculpture may speak of the fragile resonance between sun and earth (Nancy Holt), a dance may dramatize bodily movement in a way that reimagines human freedom and mobility (Pina Bausch), a novel may constitute a utopian universe, with fictional characters, stories, settings (Ursula Le Guin), and so on. Each time, I suggest, the secret is that art functions as a semidetached laboratory of imagination, allowing the elaboration of possibility in a protected space, leaving open whether and how far this could be more broadly realized or whether it is its very artificiality that allows this event to happen. The primary event of art, what Heidegger would call its work, lies in its interruption of homogenous space with its own declaration of independence. This surely has to happen for the artist, and for the audience/critic, albeit in different ways. The artist may inhabit or have insistent inklings of this new space, while our delight in the work may well take the form of a shock. The second wave comes in our recognition that art plays out, writ large, ways of shaping, projecting, and imagining highly analogous to those in the wider world, which is not only educative but also nicely questions the very distinction between art and reality.

Finally, we have earth or Earth (from background to a sustaining condition, even Gaia). We can and often do objectify the earth—think of the Apollo photographs, surveying, or road-building. Or we treat it, explicitly or tacitly, as useful material, whose very being is exhausted by its exploitability. This is the sinister side of creationism in that it can provide a ground for that tragic error.[27] Many people are deeply resistant to treating the earth as anything other than a silent supplier of all our needs, optionally created for just such an end and infinitely forgiving of our ingratitude. This is to spell out what may not actually be raised to consciousness. It is not incompatible with taking pleasure in its bounty—enjoying butterflies, birdsong, and a crisp dawn. The earth here is *essentially* a benign and untroubled, if sometimes fickle, sustaining background. The idea that the earth might be thought of as Gaia, and dying, or that its role in making human life possible might itself be threatened, or that there might be limits to its homeostatic ability to regulate the atmosphere, would be a dramatic shift, akin to the huge sense of vulnerability that Americans and others felt after 9/11. No more taken-for-granted security. On the new view, the earth is a finite, fragile ecosystem that we are changing in ways that threaten the shape of its life-sustaining capacity. This shift can come as a reasoned conclusion after reading the evidence, but it also gets registered as a dramatic transformation of the dynamic shape of our relationship with the planet. Or, in the case of 9/11, with the "outside world." Yet another analogy—consider Heidegger's insistence that the problem of the "external" world (in Hume, in Kant) is a pseudo-problem, one that rests on a fabricated dualism that we need to set aside. As Dasein, as being-in-the-world, we are always already "out there," engaged, connected. Again, the topological shift this implies is one that transforms our maps and can register as a reformatting event. For the later Heidegger, such a shift would be named by his renewed understanding of a thing as suspended within the fourfold: earth, sky, mortals, and gods. Instead of focusing on something in the center of the field, we come to understand it as distributed in four dimensions of significance—again as having its being outside itself.[28]

A fuller account of these possibilities of a transformed standpoint would include the experience of flying above the earth in a plane, which both objectifies the earth as a distant image and gives one a new sense of its wholeness. A version of such a double experience occurred for many of us when we saw images from space, first of the blue planet, then of the increasingly brown planet, where brown indicated loss of tree cover, drought, and so forth. In this terrestrial "mirror stage,"[29] we grasp the earth as a whole in the form of an image, but an image that speaks of a troubling trend. There are reports from people who spend their weekends exploring caves of a transformation in their sense of dwelling on the earth. They have not just acquired new information. Changes in physical perspective trail with them broader reflective implications, from the "*pensée du survol*" (high-flown thinking) to "notes from underground."

Taken singly, each of these examples illustrates the power of deep reversal and transformation that experience can bring. But if they are brought together, their specific relevance to bringing about "the change we need" on the environmental front becomes even clearer. The broken hammer reminds us of how much we take a fallible technology for granted. Replace the hammer with the internet in the example, and our world would collapse. The case of language is more complex. "Don't mess with Texas" was a successful anti-litter advertising slogan, manipulating public sentiment through an instrumental use of language. Heidegger's lesson, however, is that at a deeper level the instrumental understanding of language is allied with technological mastery of the earth. Letting language speak poetically opens us up to noninstrumental, nonexploitative ways of being. Musing on mortality is equally open to a range of dramatic transformations. "If we are all doomed, let's party" would not be good news for the planet. And suicide bombers may perversely have found a powerful way to rescue their lives from insignificance. Nonetheless, awareness of one's mortality, and the ultimate price being paid for our being alive by other species, can give urgency and intensity to the search for sustainable existence. Love, empathy, engagement with nonhuman animals, and the experience of art all take us out of ourselves, or mean and narrow versions of ourselves, and allow us to imagine other, less defensive, more generous, guiding dispositions. Finally coming to see the earth as itself vulnerable and fragile, profoundly disturbing to our careless innocence as it is, can shake us out of our infantile disregard for its own requirements. These various experiential reversals make possible a concerted change in how we live, move, and have our being. It is important to note here that what I am calling "experiences" here are not just a motley crew, but that the very idea of experience needs to be handled carefully. It is perhaps unavoidable that we give experience, in some sense, an evidential privilege. And yet experience is mediated through changing forms and technologies of representation (images of the earth from space), not to mention reigning narratives and language itself. The structures of reversal and transformation we have described may have a common power across the board, but it would be wrong to suppose that they could ever move us straightforwardly from the cave into the sunlight.

Transformation and the Middle Voice

Long ago in another stab at a formula for freedom, Sartre wrote of our always being able to make something of what has been made of us.[30] Deleuze writes of being worthy of the event; Derrida of awaiting the messianic, of being open to the impossible; and Heidegger says that man (truly) speaks only when he listens to the voice of language. At least in the last three cases, what is being articulated speaks with or from a middle voice that is neither straightforwardly active or passive.

This is a locus that tries the patience of more active instrumental thinkers. Surely, there is a place for clean, unproblematic agency, not least when what is involved is rule-governed calculable action with little, if any, need or opportunity for judgment. And what is to be said about habit, which seems halfway to sleepwalking? Is there not some sort of hierarchy of responsibility we are being offered here, with middle voice at the apex?

I would suggest that the possibility of these middle voice cases rests on a general character of experience, a certain instability in its structuration. Habitual action, however, puts this idea to the test. Habit seems precisely unreflective, everyday, banal. This matters because it is our consumption habits, so normalized and hardly visible as such, that arguably threaten our continuation as a species.

Habit Dehabituated

We touched on this in the Introduction, but let us briefly return to the subject of habit.

1. Habitual action is open to interruption. You may find a fly in your soup. And if such openness is constitutive, habit would already be essentially lodged in a space wider than itself.

2. This point is strengthened when we reflect that such interruption often provokes resistance, suggesting an investment in routine that may not be evident on the surface.

3. Many different levels and kinds of habits are the vital condition for any more creative activity. Poets need a normal habitual mastery of a natural language before the muse alights. Venturing into unexplored jungle requires mastery of the mechanics of walking.

4. Specifically in relation to the middle voice, habit is surprisingly fertile. Consider the experience of driving on the highway. You suddenly notice you have driven for miles without noticing what you were doing. Perhaps you were thinking about a philosophical problem, global warming, or a delightful friend. So who was driving? My sedimented driving body, one capable of all sorts of judgments (steering, accelerating, etc.), what Merleau-Ponty called operative intentionality, without the "I think" ever being called for. Or con-

sider the case in which a stranger asks for directions on the street. You switch into local informant mode, as if taking a coat off the peg. It is your town, and you are happy to help. And yet your agency is tied to a role in which you are only minimally distinguishable from other people on the street.

Each of these cases (in #4) represents a kind of simulation of the middle-voiced condition. One's autonomy is inextricably entangled with somatic and then social heteronomy, the space of Heidegger's *Das Man*. These examples suggest that if indeed our more disclosive experience is predicated on habit, we should not underestimate the prospects for a destabilizing probing of habit itself.

Where does habit fit in all this? We cannot do without all kinds of levels of habits and routines—physiological (breathing), physical (walking), patterns of thinking, forms of social life, many of which are suboptimal examples of necessary habits.

The challenge to this can come from either end—the claim that we can learn little from philosophy except to understand the necessity of how things are (Spinoza) (although that lesson may have powerful implications) and the idea that habituality itself can be transformed. The everyday can be a meditative practice. This latter idea can slice in different ways. One could choose to cultivate a meditation-friendly lifestyle. Some jobs (slaughterhouse worker, policeman, tax inspector, accountant) may be inherently resistant to such a transformation. Others may be more amenable.

The continuing attraction of existential thinking is that it is open to the possibility of such excessant experiences, those that overflow their own boundaries of possibility.[31] How can that happen? Should we be talking about an essential passivity? Or exposure?

The Instability of Experience

What would be meant by a general instability in the structuration of experience? I am taking for granted that in broad terms phenomenology captures the shape of what we ordinarily take our experience to be when we reflect on it. There are subjects and objects, acts and contents, noetic and noematic dimensions, and temporal synthesis —intentional relations accessible by a certain epoche of our natural attitude. Although Nietzsche would smell a rat here, this story is one that reflects the structure of at least "our" language, one in which we say "I"; deploy transitive and intransitive verbs, tenses, grammatical objects; and so on.

Of course, the phenomenological picture is not quite so simple. Husserl will come to speak of passive synthesis,[32] and Merleau-Ponty will famously introduce the structure of the chiasm to capture the experience of the crossing over from touching to being touched, made possible, he will insist, by the commonality of flesh, material exposure.[33] Indeed, it may be that one could work through *flesh* in this sense to explore the ground for a more general instability of experience. But I will take a different tack.

Empiricist accounts of the subject (think of Hume) doubt its substantial reality. Perhaps it is nothing more than custom and habit. At the other extreme, guarantees of identity, continuity, and coherence are provided by a transcendental ego or, in a more full-bodied way, by personhood. Arguably, the most plausible reading of this situation is that the subject is a "real" idealization, and not just on the part of philosophers but on the part of the beings that we are. There may be a cost to a certain simplifying distortion, but there are also benefits for a being trying to negotiate an extremely complex reality. We would then have to construe ourselves as dynamic physical processes capable of generating such an idealization. And we need to stir in to such an account the thought that this idealization is an operative fiction, not some epiphenomenal ghost. Without being too gestural here, the ingredients for such a story can be found in thinkers as different as Freud and Deleuze, both prefigured by Nietzsche.

If such a story is plausible, and it does helpfully address conflicting theoretical demands, it also serves to underpin what we might call the fragility or vulnerability of our experience. On this model, what we call "subject" and "object," "active" and "passive," "fore-

ground" and "background," "surface" and "depth," and so on would
be like the local scaffolding of phases of experience. Some such scaffolding might indeed be necessary for experience as such to take
place. And if the dominance of this or that model should not be
thought to be a matter of chance (sovereignty would be an example
of a recurrent set of such structures) nor is it guaranteed, either tran-
scendentally or biopolitically. On this model, these structurations
of experience are constitutive operative idealizations, positioned
by and positioning shapes of subjectivity, agency, embodiment, and
dwelling, held in place by habit, pressures of various sorts, our material circumstances (scarcity, poverty), the richness and elasticity of
the symbolic culture in which they are embedded, and so on. This
does not mean that they can be changed just by reflective exposure.
But it does mean that they are open to being disrupted, interrupted,
reversed, transformed, and displaced. In that process, something of
the shape of the scaffolding gets exposed. And other possibilities at
least get taken for a stroll.

I have distinguished, or at least suggested a distinction, between
spontaneous involuntary experiences of reversal and those that may
be cultivated. To those who would seek to encourage (or suppress)
such experiences, it matters that this distinction is not as simple as
it looks: indeed it rests on one of the very distinctions whose purity
is under scrutiny. It is important that we do not need to will these
experiences, but we can cultivate the disposition that would welcome them. Descartes, speaking of proof in both geometry and logic,
tells of running through the sequences faster and faster until they
fall under a single intuition. I used to think of this as shameful concession to the weakness of our intuition muscles, but I now think
there is something to it. When I looked at that elephant's foot, the
connections that dawned on me flooded in effortlessly and involuntarily. The conclusion I draw from this is that reflective agency can
build on involuntary (passive) responses to disclose connections and
shapes of feeling that then become available to spontaneous affectivity. And this is a metastasizing disposition, applicable with growing facility to analogous situations. Negatively, these analogized

intuitions are part of the argument against hunting animals, which can train a wider desensitization to doing violence.

I want, then, to propose a way of understanding what I take Deleuze and Derrida to share with respect to welcoming the event, the impossible. I understand the event as at least exemplified, if not exhausted, by the kinds of experiences we have been looking at, experiences that break out of the frameworks in which their ingredients originally or typically appear. To be worthy of the event, then, or to be open to it, is to attempt to develop dispositions that do justice to what these events reveal. The impossible is that which is unthinkable within some normal frame of reference, to which we typically have not worked out work-a-day responses. Experience might be thought of as a constant struggle between exposure and control. We are constantly trying to make sense, give shape, and deal with what we encounter.

But here I need to make a confession. I have always taken for granted normative privilege to a certain openness in philosophy. I take seriously Reich's and Adorno's accounts of the rigid authoritarian personality, Deleuze's and Guattari's understanding of fascistic thinking (*Anti-Oedipus*), and so on. I like to think of evil as ignorance and misunderstanding. Grasping the sources of one's own folly is not enough, but it can help cultivate personal and political dispositions that make a difference. On this view, the point of philosophy is selectively to cultivate these dispositions. Before and beyond philosophy, I am thinking of attentiveness, humility, respect, experimentation, recognition, articulation, openness, generosity, patience, engagement, creativity, imagination, and thinking. These get developed into specific philosophical virtues: critique, skepticism, reflection, close reading, dialogue, and so on. This is what lay behind my account of deconstruction as a disposition (see Chapter 4).

Given these assumptions, a responsive subject, a subject not committed to unidirectional power and control, a subject in touch with its own constitutive conditions of possibility, is a worthy goal, something to be worked toward. And if the kinds of experiences I have been discussing are valuable, it is because attending to them, cultivating them, and reflecting on them will tend to realize that end. Why is that end valuable? Can we further ground the normative privilege we are ascribing to a certain fluent transparency of being? That very formulation might seem blind to the material conditions that sustain and constrain the changes we need. But this would be a misreading. The claim is that changes in our self-understanding may well be necessary, if they are never sufficient, and that articulating these changes can help realize them. At the existential level, one would have to spell out further how it is that necessary structures of self (e.g., the need for boundaries) get colonized by the darker forces that play on fear, insecurity, and anger. If these forces do supply what is needed they do so at a tragically inflated cost. We are in the area of what Marcuse called "surplus repression."[34] But eco-ontologically, as one might say, the real prize would be release from our desire for metaphysical mastery of the earth so that it might be reinhabited.[35]

CHAPTER
7

Touched by Touching:
Toward a Carnal Hermeneutics

Weren't you asking, even before the beginning, whether we could caress or stroke each other with our eyes? And touch the look that touches you?
—*Jacques Derrida*[1]

I understand hermeneutics to be the task of interpreting plural meaning in response to the polysemy of language and life.
—*Richard Kearney*[2]

After years in the deconstructive trenches, I have recently revived an earlier affection for a certain strain of Wittgensteinian practice. Might not the point of philosophy be to encourage and inculcate dispositions that explore the constitutive complexity of often quite ordinary experience?[3] I propose here to take some concrete instances as occasions for focused reflection, openings for imagining a broader practice of carnal hermeneutics. These instances include snatches of conversation, experiences, and works of art. The general assumption is that these cases exhibit many "thinks at a time," that philosophical reflection can bring this out, and that such reflection feeds back into deepening our original experience.

"They Are Feral Cats—They Will Never Let You Touch Them"

Derrida is standing naked in his bathroom. Mary, his female cat, is looking at him, taking in, it seems, his private parts. He cannot know what the cat sees. He cannot appropriate the gaze of the feline. Yet he feels shame.[4] This feels like a twist on Sartre's account of Medusa's look, which would turn one to stone. A strange action at a distance. But how much of this story is tied up with the eyes? What if the eyes themselves were touched? We alluded to the sun in Chapter 6, taking in the fact that these very eyes that can see the sun are children of the sun. Without its heat, there is no life. Without its light, there is no vision, no eyes. The flesh of the world, as Merleau-Ponty might say. Look directly at the sun and your retina will be burned, touched by fire. But are your eyes not already touched, shaped indeed, by what they seem merely to be looking at? What happens if Derrida were actually to *touch* his cat, or be touched by it?

I had responded to an ad for mousers, feral cats taken off the streets, neutered, given their shots, and then farmed out to people with barns who need feline death squads to deal with mice. I took three siblings. I was warned: "They are feral cats—they will never let you touch them."[5] Not that they would mistake me for a mouse. It is, rather, that I should not mistake them for warm furry pets, no matter how endearing they might look. They are natural-born killers, not cute kittens. Look them in the eye, and you will see black holes from which light does not return. Four-legged psychopaths. Mice in the barn will be a thing of the past.

But they did need feeding, as mice were thin on the ground. Over time, they got used to the routine and crowded the bowl. They did not bite the hand that fed them—in time, they touched it. The hand made innocent stroking gestures. The next day, the boldest cat was coiling herself around my neck, purrcrazy. The rest is history. Perhaps a feline trauma survivor would never let me touch her. But for these cats, touch was an intoxicating magic waiting to happen. What do we learn here?

Wittgenstein says that if a lion could speak, we could not understand him. Nuzzling, purring, coiling, licking—are these not speaking? What is it about her rolling over on her back to expose soft fur
to stroking that I don't understand? If I say I know what it is like to
be a cat, isn't that an anthropocentric presumption? Doesn't a cat
live in a different world, if it has a world at all? Or do we not meet,
if meeting is ever at all possible, precisely at this carnal level? Suppose the cat licks my finger with a gentle rasping action. My pleasure comes from the edge of roughness of its tongue (a lion's tongue
would probably take off my skin) as well as from the wet warmth,
the recognition of the cat's pleasure, and the pulsing rhythm of the
licking. I am aware that the cat may be in regressive mode, revert-
ing to its time as a kitten. But are not such memories locked up in
each of us, hopefully available for deployment at the appropriate
time? Anthropocentric presumption? How about carnally crossing
the mammalian bridge? We may be human, as we say, but do we not
sport other bodies, too, ones we share with cats and one, they say,
we share with reptiles? There is difference, there are strange gaps.
Even stroking is not always plain sailing with cats. An excess of
pleasure and the claws involuntarily come out and pierce the skin
through my jeans. But there are just such gaps and strangenesses in
our sensuous and sensual relations to the opposite sex, indeed to the
same sex, to children, and even to ourselves. The problem with the
charge of anthropocentrism is that it oversimplifies the anthropos,
strips us of our layerings and differentiations. Man is a species that
is not one.

What can we learn from such an encounter with a cat? Before language, within language (I am thinking of Kristeva's semiotic here),
there is rhythm, pulsation, touch, difference, and perhaps even desire of a sort. This is true whether or not these domestic lions can
speak. The touch of my cat does not conform or fail to conform
to Heideggerian prescriptions about world disclosure. She touches
off questions about the categories I would like to think were more
stable—about species distinctness, sexuality, the erotic, language,
and indeed touch and the carnal. But, to make sure we are not fall-

ing into a trope of political correctness, she and her two brothers also confirm suspicions about multistranded commonalities, not just differences.

"My Breasts Are Too Small"

[T]he appearance alters, and from being obscure, small, and faint, grows clear, large, and vigorous.

—Berkeley[6]

The magic and the most powerful effect of women is, in philosophical language, action at a distance, "actio in distans," but this requires first of all and above all—distance.

—Nietzsche[7]

You say your breasts are too small. What are the hermeneutic stakes here? You confess, lament, admit, protest, worry, and disclaim that your breasts are too small. Do you mean as eye candy? To the touch? Too small for what? For whom? Where did you get this idea from? Why do you say this to me? For words of reassurance, consolation, and admiration? Anyway, I am touched.

What are the hermeneutic stakes here? This real-world example captures in words the interweaving of the multiple dimensions in which we find ourselves living, moving, and having our being. A woman identifies her breasts as sites of anxiety with respect to her desirability to a man or to men, despite having successfully nourished and pleasured an infant. She projects onto her lover, or prospective lover, a culturally specific preference for larger breasts. She may be right about that. But does this preference apply to her lover? And where would this general preference come from? Are we (men?) driven by visual images in their own right? Or do we associate the visual with the tactile? Does the man who sees the larger breast imagine a tactile corollary significantly more strokable, or kissable? Is this anticipated pleasure *just* a memory of the suckling delights of infancy? And what's with this "just"? Are adults deceived by the

illusion of renewed lactation? Or are the various promises constituted by the warmth of the flesh, the scent and taste of the skin, the
sound of the heartbeat, the vibration of the speaking chest, the safety
in being held—are these not quite enough, even without milk? If so,
is not the anxiety about size misplaced? Does he really think of her
as a bearer of breasts of a certain size, or does he want to touch and
stroke *her* breasts? Does not singularity trump size? And does he
bring lips, tongue, and hands to the scene, or *his* lips, *his* fingertips?
Is there such a thing as the pleasure of intimate touch in general, or
is it utterly dependent on the answers to all these questions? Is it
perhaps not that her breasts are too small but that our questions are
too small? Or is desire simply not to be second-guessed, explained,
justified, and so on? If she opts for surgical enhancement, is she not
just playing the odds? When she says to me that her breasts are too
small, she may be making a move in a game, soliciting a response
that would repudiate all comparisons. But is she then entrusting
her fate to language, to conversation? Will this quell anxiety or just
restage it at another level? ("Does he really mean it when he says
I'm perfect?") Can language speak or speak of the singular? Who is
speaking? Would her breasts speak like this? And when? Only when
neglected? Or even with happy, erect, flushed nipples? I look at you
and say, "Your breasts are perfect."

Two Calla Lilies on Pink

I am looking at a painting by Georgia O'Keeffe: *Two Calla Lilies on
Pink* (1928), which teeters on the edge of plant porn: waving erect
yellow stamens set off against flowing, light-pink petals. The allusions are obvious, and for some too obvious. And yet if we allow
ourselves to slow down the imaginative drama of looking at such
a painting, there is much to be said. At one level, only a minimal
analogical transfer is needed. For these are indeed the sexual parts
of the lily. But consider the reflective experience this opens up. Both
male and female "parts" are depicted. Does it matter whether the
viewer is a man or a woman? Do we understand ourselves as gender

positioned? Does it matter that the artist is a woman? How does it affect our experience that there are two phallic stamens? What does it mean that we can be aroused, or we can imagine being aroused, by a painting of a plant? Is the painting doing something that the actual lilies could not do? After all, there is a stylization in the painting that could not be there in the flower. The petals are like the flowing lines of a dress. Or we could say that the painting bears witness to graceful curves, lines that give on to interiority, a topology of surface and depth, which teaches us something of the fundamental shapes of tactile desire. They solicit our imaginative touch. This raises another thought—that there is something missing from this painting: the pollinating insect, the bee. Yet if it is missing from the painting, it is not missing from the experience of the painting. For it is precisely positioning us, the viewer. It is then worth noticing that the pollinating insect does its work indirectly, attracted by nectar and moving pollen around on its feet. At the heart of sexuality is something not essentially sexual at all. The touching degree zero of material transfer. At the heart of meaning, an abyss?

For Heidegger, "only because the 'senses' belong ontologically to a being which has the kind of being attuned to being-in-the-world, can they be 'touched' and 'have a sense' for something so that what touches them shows itself in an affect."[8] There would be no resistance if things did not matter to us. For Heidegger, the chair cannot "touch" the wall because it is not a site of disclosure. But what if mattering itself had its dark roots in matter?

Is this a carnal bridge too far, or does it (merely) take us further into the territory already opened up by stroking the cat? If sexuality is something at some level we share with plants, does not that fact make sexuality all the more puzzling? What would it be to understand it better? Could that involve accepting limits on understanding? And what kinds of limits would they be? Is it that whatever else, our sexual being is a incompletely thematizable ground, driving us in ways we cannot wholly explain and accounting for our existence and the shape of our dwelling in the first place? (Dependent on and typically growing up with more than one parent.)

The Pleasure of the Text

A carnal hermeneutics of language can perhaps learn something from Roland Barthes's *The Pleasure of the Text* (1975): "Perhaps for the first time in the history of criticism . . . not only a poetics of reading . . . but a much more difficult achievement, an erotics of reading. . . . Like filings which gather to form a figure in a magnetic field, the parts and pieces here do come together, determined to affirm the pleasure we must take in our reading as against the indifference of (mere) knowledge."[9] In this book, Barthes argues that writerly texts (as opposed to readerly texts, which give pleasure) give bliss (orgasm, *jouissance*) by exploding the reader's sovereignty and breaking the codes. Consider, on the one hand, Pablo Neruda and, on the other, Gerard Manley Hopkins.

I am reading a thoroughly readerly poem by Chilean poet Pablo Neruda, "Carnal Apple, Woman Filled, Burning Moon":

> Full woman, carnal apple, hot moon,
> Thick smell of seaweed, mud and light entwined.
> What dark clarity opens between your columns?
> What ancient night does he touch with your senses?

> Oh, love is a journey of water and stars,
> Of suffocating air, and brusque storms of flour:
> Love is a battle of lightning
> And two bodies—lost by a single drop of honey.

> Kiss by kiss I travel your little infinity,
> Your margins, your rivers, your tiny villages,
> And the genital fire transforms, delicious,

> Running through the narrow streets of blood,
> Until pouring out as a carnation at night,
> And being and not being is but a flicker of shade.[10]

I cannot judge the Spanish original, but Kline's English translation sparkles. Neruda is clearly writing from a male perspective, even as he embraces woman. But the play of sound, sense, and the sensuous transcends that specification. My hands assure me that stroking and caressing give shape to skin with the pulse of primal rhythm, that we are as far away from language as we could possibly be. Stop talking, they say; just let me touch you. Words melt away.

But an erotic poem like this gives the lie to this thought. Even the fingers light up in a new way. Images, allusions, and the sounds, shapes, and rhythms of words create a simulacrum of an erotic encounter, and themselves touch and move the reader. Paint the bird with a fine enough brush, and it will sing. Even fly away. Where words break in, new things may be.

What are we that we can be touched by words? Neruda's imagery is of the ocean, the sky, and then the countryside and streets. The charge of these images comes in the ways they animate without objectifying the female body. We know how a landscape can be erotically charged, how we can caress far off hills with an outstretched hand. Here the compliment is being returned, fully charged. Carried by words. Words of movement and flow that already mark the ways in which things melt in the heat of passion: "entwined," "opens," "touch," "journey," "battle," "travel," "transform," "running," "pouring out," and "flicker". And the flicker happens between being and nonbeing, as if ontology, too, is being melted by love.

Pursuing Roland Barthes's readerly/writerly distinction, Hopkins is a writerly poet. "Pied Beauty"[11] is an ode to difference, what cannot be captured in language: "All things counter, original, spare, strange." At the same time, he reinvents English by dipping back into the layer of sound and rhythm from which language is born. The writerly induces a disruption in the reader's relation to language, taking him or her out of himself or herself, opening up something akin to what Heidegger calls "listening to the voice of language." What is the "I" that is touched, moved, by Hopkins? It is the "I" at home with the middle voice, or with a certain creative dwelling at the margin of words. It is essentially liquid, and embodied.

More generally, who or what are we that we can be moved by words? A carnal hermeneutics would find ever new ways of showing how the imagination inhabits our bodies, from the pores of our skin to the ways we schematize our dynamic corporeality and our engagements with others. The erotic spawns some of the most telling ways, but there is no place for correctness here. The flesh is equally a site of lawless excitation and incitement—pain as well as pleasure, excess, and violence. If it has a transcendental face, a carnal hermeneutics would ask the question: How is all this possible? Perhaps taking a cue from Freud, it would ask about the drive to destruction, death, security, and release from stress (even anaesthesia) in addition to the search for pleasure, Thanatos as well as Eros. And all that lies in between.

"Vampire Hermeneuticks"

Half naked in the airport bathroom at Alice Springs in Australia, I found myself staring down in horror at my private parts, where it seemed as if black death was erupting out of my skin. It seemed like an unfortunate place to die without the specialized medical treatment I would surely need; I was a long way from home. I inspected the black streaks of the emerging scourge more closely. You will appreciate that I would have given almost anything to be able to substitute shame at being consumed by the gaze of Derrida's cat. At least the cat was over there, witnessing at a distance, perhaps carnivorous in Derrida's extended sense, and yet not literally devouring me. But the alien force that had invaded me here in this bathroom did not seem to brook a negotiated settlement.

Here are some facts about leeches. Leeches are hermaphrodites, each leech being both male and female, though they need each other to reproduce. Only a few species of leech feed on blood. They all have thirty-four body segments with a sucker at the front, surrounding the mouth, and another at the rear end, and many tiny eyes, which help them find food. Blood-sucking leeches can ingest several times their empty body weight in a sitting and live on a single meal for several months. I had been swimming in one of the pools in the otherwise arid Kings Canyon, situated in Watarrka National Park. Two of these little guys had found their summer lunch, and it was me.

So it was not black death breaking out of my private parts, but a couple of dark leeches who had become quite attached to me. Derrida writes that the only responsible decision requires that we go through the undecidable. Fortunately, he added that that did not mean we necessarily had time to kill before deciding. Sometimes, one needs to decide straight away, ideally yesterday. I did not hesitate. The leeches peeled off quite easily, and very responsibly, leaving a pink footprint similar in color to Georgia O'Keeffe's calla lilies where they had gently dissolved surface skin to get a better grip.

Being touched is not always welcome. We speak of inappropriate touching, of a touch of evil, of the weak-minded being "touched," perhaps by the devil, and we avoid the touch of the leper, or the untouchable. We say "touché" in fencing or disputation for a successful hit or wound. The bite of a vampire takes the kiss a touch too far. In comparison, the mini-vampire leech is more of an irritant than a threat. And if, as I suspect, what von Uexkhüll says about ticks[12] is equally true of leeches (that their behavior is governed by something like three basic sensors—blood temperature, butyric acid from the skin, and finding a hairless site), the opportunities for a sophisticated hermeneutics of such a touch are limited. But what may be true biologically is far from true symbolically and culturally. There is little more potent personal or political slur than "bloodsucker," the parasite that drinks our vital fluids. And although the leech is rivaled in this respect by the wound-cleansing maggot, its well-known medical uses are overshadowed by our anxiety at its insouciant disregard for the integrity of our bodily boundaries. To Derrida's shame at being looked at by his cat, we would counterpose disgust, horror, and alarm. This is only compounded by ticks that bury their heads in our skin, snakes that inject their venom, and worms that live coiled in the gut. Such critters all move beyond surface touch to penetration, the gentle enabling of anticoagulants to help blood flow, toxins that paralyze, and so on. They seem to know what they're getting into from the start; they know us from the inside. These creatures touch us, but the hermeneutics begins with the reversal in which we reflect on the experience of being ourselves more than touched.

What for Merleau-Ponty is a chiasm of touching, primarily in the literal sense, is here being thought of as a broader possibility of reversal, in which each and every passivity can be the subject of a reflexive receptivity, either as a new complex experience or as an ongoing process of transformation.

These experiences of violent touch can, of course, be replicated in the human world, with rape, torture, hurricanes, disease, and violence of all sorts, not to mention the slings and arrows of outrageous fortune, where it is our material exposure that is at issue, an exposure constitutive of our being. Here is perhaps one of the richest sites of a carnal hermeneutics—experiences of joyful explosion of boundaries and of the anxious defense of such boundaries. Some of these boundaries are, as we say, real and others merely constructed, perhaps manufactured, the better to manipulate us. Touchings of all sorts do not merely disclose such boundaries; they open them to scrutiny, transformation, renegotiation.[13] A critical carnal hermeneutics would reflect on the emancipatory possibilities latent in the maintenance and the overcoming of the various boundaries that constitute both our identity and our delight.

PART
III

Reoccupy Earth

PART

III

Reoccupy Earth

My Place in the Sun

It is through time, especially history, that place is distinct from space. This claim survives our moving away from a naive naturalistic understanding of the past to one constructed and constituted so as to include narrative, intentions, and projections, even when these form the basis for serious contestation of what we take to be the past.

Space and Place

The distinction between space and place is not difficult to grasp. On the one hand, geometry, on the other, location, permeated one way or another with meaning. We identify space with measurement, with the neutral, objective way in which extended things are separated from each other. Space, in this way, is understood as "external," in classical contrast with mind, or "thinking stuff," which is internal, a distinction that itself sounds spatial but is not. Classically, it was Descartes who distinguished res extensa from res cogitans. This sense of the objectivity and neutrality of space is complicated somewhat both by theories of perspective and by the theory of relativity, in which space is subjected to perspectival considerations (perhaps ineliminably), suggesting that the position of the subject, orientation, is an essential ingredient, even to "external," "objective" space. But the outcomes are still calculable, even when we take into account the requirements of perspectival representation. Adding a subject just brings in a new "angle."

Place, however, seems very different. When American poet Gary Snyder talks of the need to eventually put down roots, he is making a move within a literary and cultural tradition that celebrates mobility and quest.[1] Moreover, the distinction between space and place can quickly become charged, even politically, often tied up with territorial conquest, mapping, and property. In both the United States and Australia, spaces and places were nullified, neutralized, by surveying practices that located even natural features on geometrical grids, with some residual respect given to rivers and peaks.

"Place" operates at many levels and registers. "A place for everything, and everything in its place" refers to a well-designed kitchen or workshop in which a range of useful things is easily accessed. One place (a kitchen) will provide subordinate places for smaller things. "Come back to my place," "Your place or mine?" are invitations to a home, a dwelling, a locus of personal life and meaning. To win first place in a competition is the mark of accomplishment in which one is differentially ranked in relation to others. And there are many situations in which we evoke the value of place without saying, "This is (a) place"—as when people talk about their favorite cities, restaurants, or vacation destinations. We do, it is true, speak of place in more subjective ways ("I was in a bad place for months after she left."), but this seems like a metaphorical extension of the more usual sense of a location whose meaning is shared with others. Indeed, if one had to choose one word to distinguish place from space, it would be meaning. I am emphasizing the contribution of time and history to such meaning.

At this point, one could launch into a neo-Aristotelian description of the conceptual geography of place, aided and abetted by, for example, Dilthey's hermeneutics. Or (with Ed Casey) we could creatively explore the phenomenology of place in its extraordinary variety.[2] These are both valuable exercises for another occasion.

Place as Site of Contestation

I consider here situations in which "place" is the site of contestation, and where history plays a central role in that contestation. I want to highlight and explore the connection between the existential sense of right attached to one's bodily existence, the values associated with dwelling and home, and the political and ecological consequences of the ways in which the scope of this sense of right is interpreted. Place and its history and promise are central to these issues.

Here is the argument: We each take for granted as the ground of every other right and privilege our individual right to exist, to have our "place in the sun." This is rarely contested, and it is undoubtedly defensible. We are speaking of my (mere) right to exist, not as a bloated plutocrat, one of the 1 percent, but as a mere human. The image of "my place in the sun" makes it clear that we are not talking about a privileged spot on a private beach but a place on earth that does not deprive others of what is theirs. Not a zero-sum game.[3] After the holocaust, it must have seemed to many that merely being alive was a special privilege for which one might feel guilty. Levinas's meditation on "my place in the sun" would be a good example.[4] But if (to be very quick) the conditions under which this was or seemed like a zero-sum game (my survival = your death) were the work of the devil, then one is surely buying into evil in endorsing this position, even in feeling this guilt. Nonetheless, even if one's right to exist is something like an unconditional background, it is one that can, at least apparently, be suspended or modified. The idea of an "unconditional" right may not ultimately fly. It may be that there are conditions, such as lifeboat situations, under which unconditional rights would be suspended. If there were unconditional rights, the right to the minimal means of existence, given abundance all around, seems like a strong candidate and on a quite different level to the right to free speech, education, health care, and so on. It seems to be a condition for these other rights that one continues in existence. And yet when one says "given abundance all around," that sounds like a condition. What if that were not satisfied?

And what consequences does this have when one takes seriously the idea that continued bodily existence can only artificially
be separated from those interactions with the world that provide
sustenance—what we often call labor? Is there then a right to work,
and where does that take us?

When John Locke famously connected the right to property to
whatever one had "mixed one's labor with," he included an important caveat—that the exercise of such property rights would not significantly affect the opportunity for others to do the same.[5] And we
might add, not at the time, nor subsequently. If we relegate to the
cognitive margins the existence of native inhabitants and nonhuman
creatures, this looks like a splendid way of justifying and motivating
the colonizing of the New World, and indeed they did, and it was.

The question I am posing is this: Is there a slippery slope from
distinguishing place from space in the existentially uplifting way
evoked by Gary Snyder to the attribution of a narrative value to
place that would legitimate possession, exclusive use, privilege, and
so forth? And if there is, in fact, a slippery slope, can we mark the
point at which such slippage becomes dangerous?

If one has something like a natural right to one's own corporeal
existence (embodied in the right to self-defense), the obvious next
step is whether that right implies or can properly be extended to the
locus of one's life—one's home, for example. Some sort of extension seems inevitable once we admit, as we have suggested, that
bodily existence is impossible without its relation to some sort of
sustaining ground. We need air, food, water, shelter, and space to
exercise just to preserve our biological existence. Only thereafter do
we need information, companionship, and so on.[6] Bodies are not just
things but indices of our essential relationality, both in a physical
and social sense. Ears, eyes, mouth, hands, and genitalia all shout
this in different ways. But it is not clear what implications this has
for any particular situation one might find oneself in. The burglar
who finds the beer (or Perrier) in your fridge cannot justify his drinking it on the grounds that we all need to drink. Juridically weighed
statements of our collective responsibility for providing minimal

conditions of survival even to social outcasts invented the jail cell. But no one supposes that a prisoner has a right to some particular cell (#353 at San Quentin), even if he has gotten used to it. Similar considerations apply to the provision of public housing for the poor. Once it is accepted that we often and appropriately live in groups, such as families, local sociocultural norms will generate minimal housing units—for example, apartments and small houses. In each of these situations, one's right to continuing existence implies *some* place, not some particular address. (Kant's version of this is that just by having a body, we have a right to be "somewhere" on the planet but nowhere in particular.)

These cases of social provision make visible only the most basic layers of place—physical sustenance and a locus for shared space—what we might call (a) "home."[7] Minimal though it may be, we should not underestimate its significance. The plight of the homeless, the refugee, those who lose their homes in a hurricane cannot be overestimated. In the home is centered both a whole set of utilities (from which one can venture out into the world and to which one can return) and a whole slew of meanings. Indeed, refugee camps that begin as rows of tents turn into complex communities with stores and schools.

Mixing One's Meaning

The hermeneutics of place begins, one suspects, with the body and its physical and social needs. But with considerations of meaning, it flowers, and beyond the point at which minimal needs are being satisfied. Whether or not we endorse Locke's understanding of property, there is something compelling about this account ("mixing one's labor") for understanding this blossoming of meaning. In the case of involuntary incarceration, meaning is at a premium. A cell mouse may become a prisoner's closest companion. Maintenance neglect is a common problem in social housing provision. There are limited opportunities for mixing not just labor but meaning with one's surroundings, for the investment of time, effort, energy, and planning.

Where such opportunities abound, meaning flourishes. Herein lies the attraction of the pioneer spirit—where one would find, occupy, and develop a piece of land and build a house. And, more standardly, it explains the legal protection offered (under normal conditions) by property rights. They enable long-term planning and investment, the laying down of associations and memories with a place, and so on. These are the sorts of considerations that lie at the heart of objections to people like Proudhon ("Property is theft"). But the seeds of his appeal are not hard to find.

The condition Locke laid down for legitimately connecting property to labor was that one not deprive others of the same opportunity—much as we make parallel remarks today about freedom and sustainable development. The clearest cases are those in which there is an infinite (or practically boundless) supply of the material in question. Here we applaud those who freely bestow deep, rich, complex layers of meaning on the spaces they occupy.

Locke's premise was that property is not a zero-sum game. True, if I own something, you cannot, at the same time, also be the exclusive owner (though we could share). But if creating property were like writing poetry, it would be largely benign. Admittedly, if I write a poem, you cannot write the very same poem and unproblematically claim it as yours.[8] We do not think of this as a restriction because the combinatorial possibilities of language seem infinite.

It turned out that the premise on which Locke's definition of property rested was more problematic than anticipated. When applied to the American colonies, not only were there original Native American occupiers of the land, but what was meant by "mixing one's labor" was not at all clear. Does running cattle on the land count?[9] It soon became clear that there were choice locations (springs, soil, minerals such as gold, ocean views) and a limited supply of them. On the ground, mixed in with Locke's agrocentric compatibilism, was violence and land grabbing. Moreover, if we look more broadly at the way place and property come into being, it is hard not to conclude that forcible expropriation of existing rights is the norm.

If we take this thought seriously, we would be talking not about mixing our labor with the land but mixing our blood. Or the blood of others: "enemies," sons, fellow creatures.

At this point, I want to step back by opening up a new question: Is the meaning of the structure of a place sui generis, essentially sustaining itself, or does it rest on something quite different? I have already, in effect, addressed this question in connection with those meanings associated with property, connecting them back to bodies and work. Does such a grounding demonstrate a wider truth?

Place and Territory

Consider the idea of territory. The meaning of this term ranges across the spectrum from politics, through economics, to the ecological. Britain has (or used to have) Crown territories, "possessions" it administered without their being part of the UK. Marketers and salespeople have distinct territories where they can focus their efforts without competition from their colleagues. And nonhuman creatures have territories, too, which they mark, patrol, and defend. These are clearly places, spaces that sustain and distribute meaning. But these meanings are tied back to imperatives—interests, forces, histories. Colonies were conquered and subdued for their exploitable resources (including human bodies) and their strategic location. Commercial territories are allocations that encourage the investment of energy, time, and attention by guaranteeing exclusive access and control. Animals maintain territories to control mating or to protect their food supply.

If territory is a kind of "place," its skein of meanings (significant subordinate places or sites, its history of important events, ways its boundaries are marked—local dialect, color-coded maps, urine) derives from, rests on, what we could call vital interests—profit, life/death, reproduction, and so on. Predicated on these factors, lines and boundaries arise.

Place and Contestation

If, for a moment, we took territory to be exemplary for thinking place, it would have dramatic consequences for our basic model. Mixing one's labor (or one's meaning) seems like a relation that a subject has with its world, independently of what others are doing, hence the importance of the "infinite supply of material" caveat. We may suppose that battles over special sites would be exceptions, marginal cases, but what if they were, in fact, the norm? Or that we came to see my (or our) relation to place as always in principle contestatory, even if I only want to name it. Can I cultivate a garden, or participate in your garden cultivation without excluding or helping to exclude others from making those same investments? Turning Locke on his head, what happens to mixing one's labor under conditions of serious scarcity?

In extreme situations, it is often said: "It's every man for himself!" It is not entirely clear what this means. It could mean that in the absence of the opportunity for a coherent group response, each must seek his own survival in the best way possible, even if that strategy would not, under better conditions, optimize the greatest number of survivors or indeed maximize one's individual chance of survival. Or it could mean that the moral niceties of collective life have ceased to apply; walk over your grandmother if that helps. This latter idea, in turn, could be interpreted as the breakdown of all order, or a return to the primitive "law" of survival. Such a law would justify, for example, killing the other man in a lifeboat stocked with only enough food for one to survive.

If there is some sort of natural right to ensure one's survival, even if, in extremis, under conditions of genuine scarcity, it involves taking the life of another, and if some minimal sense of property can be justified as an extension of the right to ensure one's survival, this would seem to offer a justification for the expropriation of another's property, even if they themselves needed it to survive. If I am starving, I can justify stealing from your garden (even if that would mean you would starve) or even stealing your garden. I would not need to dispute the legality of your title. I would need only to be genuinely desperate to acquire, in whatever way necessary, the means to survive.

I do not say that there is or could be a legal system that would endorse such a course of action.[10] There could be a system that would treat desperate need as a mitigating circumstance in the case of theft, even violence, for example. And that mitigation would evaporate if the situation was not, in fact, desperate, or if I could just as easily have asked for some food.

The point I am making is this: even if we cannot legally codify this, we cannot rule out being able to justify violence at the origin of possession and dispossession of property. But just as Locke's account of property through mixing one's labor rests on not depriving others of the same opportunity, so, too, even this limited justification of violence is conditional, with conditions not unlike those for a just war. You have to be genuinely desperate, have exhausted other means, and the violence must not go further than necessary. We must also take into consideration the question of time and history. If you are away from home in the winter and I survive the cold by helping myself to your garden vegetables, this is not a justification for permanently taking your whole garden. By spring, things will be looking up. Suppose, however, that you do not come back, and over the years I make your garden mine, and perhaps your house, too. I fix the leaking roof, add an extension, raise a family. Then you come back.[11]

There is an old story of a poacher caught red-handed at dawn with salmon stuffed down his trousers. The squire tells him he is trespassing, that *he* owns this land. The poacher asks him how it is that he owns it. "My forefathers fought for this land!" "Well," said the poacher, "I will fight you for it now."

I suggested earlier that the desperate man who kills for place, for a plot of land for his family to live on, could "justify" doing so, even though it might not be legal. One way of understanding this would be to say that he could tell a story, a narrative, in which his actions would be understandable, perhaps even reasonable. "What choice did he have?" But when my neighbor returns after many years to reclaim his property, he may have an equally good story. He was falsely imprisoned by a brutal dictatorship and unable to make contact. What now?

We have talked about place as a site in which labor and meaning are mixed. We have supplemented this two-term relation with a third term—the other—with whom I may have to compete for possession of this place, or even to stamp my meaning onto it, perhaps to the point of violence. I suggested at the outset that what makes place distinct from space is its temporality. Most likely temporality, especially history, is how meaning, and perhaps the sacred, gets added to the mix.

The Historical Constitution of Place

Let us now think about the historical constitution of place. I live on a large farm in Tennessee. Some four hundred million years ago, what is now limestone was the bed of the sea. Numerous coral fossils pepper the surface of wide slabs of rock, often many feet wide. In the creek not far away, Native American arrowheads and scrapers still surface, anything up to ten thousand years old. At the top of the hill, there is a Civil War installation of undetermined function from the 1860s. Google Maps shows outlines of fields long ploughed under and farming patterns dating back to 1900. Then there are derelict sheds and barns, roads overgrown and newly cut, and skulls of dead creatures, wild and domesticated. Numerous living beings with their own lives and times—from fleeting dragonflies to slower snakes. Some of this time is current—limestone dissolving in the rain, frogs splashing in the pond, sunlight shimmering in the spiderwebs. These concurrent rhythms interweave to make something like the fabric of the present. But the sedimented layers of the past are equally real and constitute a quite different dimension. We may imagine that the reality of the past of a place lies only in what remains today—fossils, blurred outlines, relics, skulls, the contours of the landscape. On this model, events that occurred "here" once upon a time are, as they say, history—done, over with, finished. It is not clear how to change the paradigm here, but what I am proposing is that its past be treated as a real part of what we understand a place to be. Then, all the events, processes, and changes that a place "witnessed" would be part of that place. A place has, as it were, temporal roots. This seems especially plausible in the case of a battlefield, a graveyard, ground zero, a historic town—where well-known events of human significance are recorded and commemorated. The more general claim is that unknown events, events in geological time, are also "part" of what constitutes a particular place. They do not need to have been recorded somewhere, but they did happen here, and so they are part of what "here" is. This is a realist view of place. Interestingly, it may supply a ground for a rather different ontology.

Near an English Civil War battlefield at Edge Hill in Warwickshire, there is a mound called Blood Copse, where, it is said, many of the casualties from that battle were buried. To this day, there are reports of ghosts of dead soldiers walking by and dogs that get too close howling at night. I do not know what to say about the dogs, but ghosts are, arguably, a visualized acknowledgment of the reality of the past on the part of those who still bear its memory, within an ontology that wants things to be present if they are to count as real. Derrida's evocation of the specter "of" Marx (Marx as specter, and Marx's evocation of the "specter of communism") makes an analogous point. To speak of a ghost or a specter is to speak of the way the dead "live on" through memory, but not just as memories. Memory gives us access, however imperfect, to their past lives. In the case of soldiers who fell tragically in a short, indecisive battle, or a great thinker who many have tried to "bury," we are dealing with what we might call unsettled memory, memory of events or people that challenge or disrupt normalizing narrative. A long discussion of trauma could begin here!

The background here is a certain realism. Places literally contain their past.[12] But our grasp of that past can generate strange phenomena. Of course, when these are repeated (e.g., ghost sightings), the growing record of such "sightings" itself becomes part of the past.

The Place of the Future

It might be thought obvious that the past would have a privilege over the future in determining place. The past consists of real sedimented layers of events, while the future is as yet undetermined. And yet it would be a mistake to cling to too simplistic an account of this privilege. Not only is the developing meaning of (a) place tied up with often contested possibilities of fulfillment of contemporary promise, both rhetorically and causally, but the same is true of the past. The layers of past events that constitute place themselves project possible futures, which may or may not have subsequently been realized. This is true both of explicit projections (plans, programs, etc.) and of more gestural indicators (things must change, enough is enough, we cannot go on like this). Many past events themselves projected temporal horizons of significance, inheriting and witnessing the past, anticipating and adumbrating the future. A certain naturalism is tempting at this point, even as it includes intentional phenomena. The *soixante-huitards* believed that the future could be shaped by imagination: "L'imagination au pouvoir!" The cobblestoned streets of Paris bear this history, with all its unrealized aspirations, as facts, even if those hopes have been largely forgotten. And as much as past projections mattered causally in shaping action, whether successfully or otherwise, the same is true today. But there remains an asymmetry. The past, we suppose, with all its complexity, actually happened. Our anticipations of the future, our desires, and our plans are all equally real and in one way or another capable of being efficacious. But if we take as our standpoint a dynamic emergent sense of place, or place as a continuous sedimentation of meaning, the future *as such* is not part of the meaning of (a) place. Hegel was right— *Wesen ist, was gewesen ist.* Being is what has been. The future in that sense is a different order of being, essentially unrealized.

Once we break with the illusion of presentism, this generous naturalism seems plausible. It is not that the past *is present*, but that, as past, it is a component of the real. It does not "live on," except in and through its effects. It does not need to. It really happened. It has not gone anywhere; it is, rather, some*when*. We may find a spatial analogy helpful, as in the celebrated analogy of the iceberg, much of which floats below the surface.[13] But this is in no way to reduce time to space. The past is the past.

I have tried to elucidate here a certain intuition about the temporally constitutive thickness of place and how that contributes to its meaning. This relies on a simplistic accommodation of the intentional within the natural. In addition to what happened, there is also what we wanted to happen, what we feared or imagined might happen. These are not posited as possible worlds but as subjective or intersubjective attitudes that can sit casually enough alongside the first-order facts.

Ontological Musings

This objectivism is, however, problematic. It is not just that we have no access to the past except through our narrative constructions. In important ways, nothing ever happens at all without being articulated one way or another. This is perhaps the point of Derrida's *Il n'y a pas de hors text* ("there is nothing outside the text"). If it seems to get close to a certain idealism, one of a transcendental flavor, it rather affirms the centrality of a certain complexity for things or events to have any distinctive identity. If something happens, if an event occurs, we are not just saying that the cosmos perdures, or continues. *Something* happens. This rather than that. It rained this morning. The rain started and it stopped. It was rain not snow. It rained this side of town not that. A mass of raindrops fell, each with a common cause (clouds). It started as a heavy downpour and later eased up.

We need an ontology that acknowledges distinctions, differences, on-the-ground, "out-there," as well as our constructive and selective activity in creating meaningful shapes on the basis of intrinsic material differentiations, and further acknowledges the difficulty of identifying those intrinsic differentiations without already putting our constitutive powers into gear. Moreover, we need to recognize that our meaning-construction activity is underdetermined by intrinsic differentiation. Think of the different rhythms one can construct as a passenger from the sounds of a train running on a track. We also need to recognize that while there are clearly opportunities for individual, personal meaning construction, *place* particularly invites shared collective synthesis through ritual, repetition, and narration.

If we accept that human meaning construction is a process of differentiation and synthesis laid over a natural world in which difference and connection are already happening (in which we are already involved and from which we cannot entirely separate ourselves), it must further be acknowledged that the extra we bring is not (just) truth in some ungrounded, acontextual sense but an articulation of interest, and interests that we might call dwelling. Implicit in this formulation, however, is first the possibility of a fracturing of the "we" and the further possibility that this fracturing might not be a mere possibility but at the heart of the matter. "We" can fracture into specific groupings of humans, set against each other. But it can also imply "we humans," set against the other creatures with whom we share the planet.

The central implication of these observations is that the naturalistic/objectivist account of the thickness of meaning intrinsic to place will not do. Does the intuitive appeal, such as it is, of the past being embedded in place depend on it? If we start to include memories, attitudes, and narratives, don't we lose the solidity that seems to be provided by the naturalistic account? Analogues to geological strata are one thing, but the swirling complexities of people's transient beliefs about what's happening are quite another.

We are now in a better position to consider place as a site of contestation. What this implies is that the articulation of meaning in and through history that gives place its depth can be expected not merely to testify to some group's (a "people") affection, depth of memory, identity investments, and so on but also to its attempt to exclude or legitimate domination over other contenders for the same space. The meanings of place are rarely free of territoriality. But that implies that the rhetoric of place is inextricable from the kind of claim to property rights we discussed previously. Developing such a relation to place (mythological, sacred, narrative) can be treated as mixing one's symbolic labor with a piece of the earth. Yet if analogous caveats and conditions obtain here, the legitimacy of such entitlement would rest either on an unlimited supply of similar sites or on dire necessity. I suggested that *some sort* of justification could be made even for killing one's neighbor if the alternative really was one's own death or that of one's family. One could "understand" someone doing that, even if one could not approve of it. There are also special new conditions, hard though they might be to specify, such as that the events being knitted together can plausibly be said to have actually happened, if they are not openly acknowledged to be myth. And that the actors in these events have some plausible connection to "us." If the inhabitants of a land weave their history together with Viking invaders, it is important both that there were Viking invaders and that "we" have some special connection with them. If other contenders equally identify themselves with the same invaders, we have a problem. The plausibility of the narrative construction will be a mix of objective factors—whether what is being claimed as a continuity meets a minimal standard of plausibility, for example, logical consistency, whether "we" want to believe it, and perhaps whether it is in our interest to believe it.

Consider two extreme cases: (1) the status of Jerusalem (2) that of "Man's Place in Nature."[14] Each of these "places" is heavily contested though in different ways.

A Tale of Three Cities

Jerusalem both personifies and symbolizes the "sanctity of place" for all religions deriving from or responding to biblical scripture.[15]

Jerusalem is claimed as a sacred site by three religions. What follows is a series of freely available nonscholarly accounts of how/why Jerusalem is important to each of them.[16]

A city divided between east-central Israel and the Israeli-occupied West Bank. Jerusalem was founded as far back as the fourth millennium B.C. and was ruled by the Canaanites, Hebrews, Greeks, Romans, Persians, Arabs, Crusaders, Turks, and British before being divided in 1949 into eastern and western sectors under Israeli and Jordanian control. In 1967, Israeli forces captured the eastern sector from Jordan, later declaring the city as a whole to be the capital of Israel. The legal status of Jerusalem, considered a holy city to Jews, Muslims, and Christians, remains fiercely disputed. Population: 729,000.[17]

Around the year 1010 B.C.E., King David defeated the Jebusites in Jerusalem and decided to make the city his administrative capital. When he brought the Ark of the Covenant to the city, he stripped the Twelve Tribes of the spiritual source of their power and concentrated it in his own hands.[18]

Jerusalem is also very important to Christianity, as Jesus Christ lived and died here. The Christian quarter alone houses some 40 religious buildings. . . . One of the most prominent and important sites . . . is the Via Dolorosa . . . Jesus' final path, which according to Christian tradition led from the courthouse to Golgotha Hill, where he was crucified and buried. . . . The Church of the Holy Sepulcher is a pilgrimage site for millions of Christians from all over the world.[19]

Jerusalem is considered a sacred site in Sunni Islamic tradition, along with Mecca and Medina. Islamic tradition holds that previous prophets were associated with the city, and that the Islamic prophet Muhammad visited the city on a nocturnal journey. Due to such significance it was the first Qibla (direction of prayer) for Muslims and the prophet Muhammad designated the Al-Aqsa for pilgrimage. . . . Muhammad is believed to have been taken by the miraculous steed Buraq to visit Jerusalem, where he prayed, and then to visit heaven, in a single night in the year 620."[20]

These accounts already anticipate ways of privileging one version over another. Length of (semi-)continuous occupation would favor the Jews. Spiritual significance seems to be shared, as well as evidence of a canny awareness of what needs to happen (to be built) to strengthen the case for distinctive legitimacy. But how can we best thematize the role of history here in constituting the meaning(s) of this place? If all meaning is differential, whether one is aware of participating in that process, Jerusalem seems to take this to another level. It seems, at least, that it is not merely the case that the status of Jerusalem is fought over now but that this very question of legitimacy dispute itself has a long history, a second-order truth that the various contenders for exclusivity might arguably agree on, even as they would disagree about the verdict. We could go further: Jerusalem appears here because it is *exemplary*, combining an extreme historical determination of meaning, *some* shared agreement about the facts, and disagreement on how to interpret them. Many of these disagreements are performative in character—reaffirming the party line. Some will be more or less disinterested. It may be, however, that Jerusalem is too good an example, to the point of not being at all typical. It may be a really good example of a continuing battleground, in which the contestation of history and the history of contestation are inextricably interwoven. It would be too good an example, because it goes beyond the claim that the history of a place is an intrinsic part of it. If every history reflects a certain interest, the history of Jerusalem, weaving together and/or counterpoising so many different interests, both confirms and belies that generalization. It is not clear, for example, how to count the Jerusalems. One? Three? But even if Jerusalem is "too good" an example, it nonetheless bears witness, albeit problematically and in excess, to the historical constitution of place.

Kant, Sharing the Earth and the "Right to be Somewhere"

What, then, of the earth itself? Can I not demand some minor version of my place in the sun as a natural right? Kant argues something very much like this when he speaks (in his *Doctrine of Right*) of our common ownership of the earth's surface and our right to be "somewhere" just by virtue of being born.[21] This doesn't provide a right to be anywhere in particular (see our earlier discussion), but it does afford a right not to be turned away when you find yourself stranded through no fault of your own (e.g., shipwrecked). The argument stems, parallel to our earlier supposition, from one's mere embodiment, which gives one the right to the space occupied by one's own body and brings one into community with others with whom one unavoidably shares the earth as a finite, bounded territory. While this right to be somewhere does not specify how that is to be concretely realized, as Jakob Huber convincingly shows, it rather provides a framework, or guidance, for negotiations with others in the same position.[22]

The earth, then, is a place not simply in a spatial sense—though its being bounded and finite is important—but in a civil sense, for we necessarily share it with other similarly embodied beings. We all breathe the same air, for example, which perhaps lies behind the peculiar horror attached to chemical (gas) warfare and to the gas chambers. Kant was far from blind to the importance of being able to give teeth to this "right to be somewhere" and for this to precede formal property rights. He was highly critical of the colonial expropriation of land that ignored the fact of settled occupancy. And he stresses the importance of limited international hospitality in a world increasingly based on trade. There are three pressing challenges to thinking through our common possession of the earth that were less obvious in Kant's time: first, the rights of future humans, whose "right to be somewhere" is problematically predicated on their being born; second, the possibility of planetary colonization, which requires us either to expand our sense of "earth" or to change our thinking quite fundamentally; and third, whether and how nonhumans share in this common ownership of the earth, on this model. As embodied beings breathing the same air, it is hard to see how they can be excluded from consideration. Kant's blend of natural and civil senses of sharing the earth surely starts to crumble here. For if nonhumans cannot negotiate as moral agents, accepting duties as well as bearing rights, they clearly have needs and interests, homes and territories that would surely align them with the indigenous victims of colonial exploitation and land expropriation that Kant deplores. Hegel complained that Kant's moral philosophy was too abstract and lacking in real-world traction. But this is perhaps an advantage when dealing with the intractable complexity of an emerging concrete situation, where such guidance may serve us better than any specific rule.

Man's Place in Nature

Writing of "Man's Place in Nature" challenges the "we" in a different way. At the heart of the phrase is some sort of distinction between man and nature, even as we seek to bring the two back into alignment. The idea of evolution, promulgated by Darwin in 1859, only four years before Huxley's book of this name, inserted man back into nature through our biological *history* but also reasserted our species privilege as the latest achievement of that process. The word "place" does not function as the name of a literal space, but the way it does function confirms the normative sense we have been arguing for, made explicit in such a phrase as "first place." To have a place is to belong, somewhere at least. (Cf. "A place for everything and everything in its place.") To speak of man's place in nature is to reaffirm that we *have* a place in the face of the death of God and the broader displacements of secularization. Contestation is implicit here. What is at stake is whether we (still) have a place, and if so, where is it located, and how is it to be conceived? In light of our earlier remarks about the privilege of the past over the future, it is interesting to note that Teilhard de Chardin extends his own brand of evolutionary thinking into the future. Spirituality is man's distinctive contribution to evolution, and it opens up a future in which that spirituality will be developed and exercised. This is quite compatible with our reservations about how the future defines place in the ordinary sense. Projections (current and past) of possible futures properly play a role in our understanding of who and where we are. Such projections embody and fulfill the very normative understandings of man that can anchor our sense of having a place (in nature). It is not implausible, for example, that developing a certain selfless wisdom will make it possible for us to rescue the planet, and by extension ourselves, from the consequences of a myopic and inadequately evolved model of agency. Continuing to externalize the toxic and dystopic by-products of our seemingly productive activities is unsustainable. Recognizing this, and projecting a different path, is necessary for the changes we need to happen in a managed way. We do not have to subscribe to Teilhard's spiritual language to acknowledge a certain aporetic dimension to our relation to nature, as when we may be said to be in the world but not entirely of the world. Perhaps Heidegger is echoing this when he recommends at least a soupçon of the uncanny as a condition of being truly "at home" in the world. Our proper place would be to be slightly displaced.

Conclusion: Our Place in the Sun

The sun bathes the earth in light and warmth. There would seem to be more than enough to go around. Having such a place does not seem like a zero-sum game. Yet we battle for the well-lit room in the house, the best spot on the beach, and the ocean view, not to mention the shady spot in a summer parking lot. To be placed in a race is very much a zero-sum game. We have argued throughout that while the intuitive plausibility of history being essential to place rests on a somewhat naive naturalistic grasp of that history as including everything that ever happened here, the thesis survives the inclusion of intentional attitudes and narrative constructions, even contradictory and conflicting ones, albeit in a more problematic (and interesting) form.

9

On Being Haunted by the Future

We propose to speak of a democracy to come, not of a future democracy in the future present, not even of a regulating idea, in the Kantian sense, or of a utopia—at least to the extent that their inaccessibility would still retain the temporal form of a future present, of a future modality of the living present.[1]
—*Derrida*

The accumulation of greenhouse gases in the atmosphere has now crossed a threshold, set down by scientists from around the world at a conference in Britain, beyond which really dangerous climate change is likely to be unstoppable.[2]
— *Michael McCarthy*

The future is not what it was. Once upon a time, we could be forgiven our hopes for better times. Now, it would seem, we are a sadder and wiser species, chastened by our success in demonstrating just how dangerous our experiments in social engineering can be. Great leaps forward have landed us many steps back. And in the process, some of our most deeply held values—community, security, and freedom—have become shamelessly abused and despoiled.

And yet, for there to be hope at all, indeed for there to be life and experience here and now, we have no alternative but to go back to the future and relate to it differently. It may not best operate as a legitimating horizon for projective planning, but a new dispensation is called for, a new way of figuring the future.

Heidegger tells us that we can choose between going counter to a great thinker or going to their encounter. Derrida writes of the need to combine faithfulness and betrayal in responding to a legacy, that we have to reinvent in order to take forward. I attempt here to go to Derrida's encounter with time and the future, and at least propose a certain *Verwindung*, a twisting free, or transformative reworking, one that gives greater recognition to the ethical, political, and more broadly philosophical puzzles and paradoxes that arise when we focus not just on the unpredictable and surprising event but on our apparent inability to deal with the all-too-predictable.

In the Beginning

In his early writings, Derrida takes over, generalizes, and streamlines the critique of the value or ideal of presence already initiated by Heidegger, stripping it of any residual nostalgia for the unpresentable. Besides Heidegger's work, in his seminal and oft-quoted essay "Differance," Derrida assembles other confluent currents that we recognize under the names of Hegel, Freud, Nietzsche, Levinas, and Saussure.[3] In each case, what is being emphasized is the dependence of any unity or identity on various dimensions of differentiation and deferral, that is, both "spatial" and "temporal" difference. In each case, he insists that we are not dealing with a local or temporary absence that could be remedied elsewhere or at a later time. Rather, we are dealing with a structural, quasi-transcendental condition that makes possible (and, strictly speaking, impossible) the value and experience of presence. Fullness of meaning, what Husserl called meaning-fulfillment, is permanently deferred, even as the lure of its possibility continues to attract us and draw us forward. There are three ways in which Heidegger's earlier treatment of time prepares the way for Derrida here. Drawing on Husserl's phenomenological treatment of time consciousness in terms of protention and retention, Heidegger offers us, in *Being and Time*, an account of the interweaving of the three temporal ecstases we commonly refer to as past, present, and future, now understood as existential dispositions (e.g., being-as-having-been). The present becomes unthinkable except as overlaid, traced-through by a past that haunts it, and the horizon of a future on which possibilities are projected. And it is through the theme of possibility (and impossibility) in the shape of his discussion of authenticity (choosing my ownmost possibility) and being-toward-death (the possibility of my impossibility) that Heidegger comes to give a privilege to the future over the other two ecstases. This privilege does not, however, much outlast *Being and Time*. For the privilege rests on the "ontological" work being done by these strange "possibilities." Death, in particular, works to disappropriate any clinging to a substantial sense of self. And we may suppose that it is such reflections that provide a gateway from the more existential orientation of *Being and Time* to his later writing. Looking back, we can see both Husserl's turn toward consciousness and Heidegger's turn to the ontological interrogation inherent in human existence as ways in which we first move away from the vulgar sense of time (toward temporality). In doing so, however, we start to understand the ways in which the very idea of time or temporality may itself mark the site of a profound struggle even over the privilege of being or the ontological. This struggle has already begun in Heidegger's own work—in his repudiation of the language of ontology, his crossing out of being, and his movement from philosophy to thinking. It is continued by Levinas and Derrida in different ways.

The point of this brief review is that with Derrida, as with Heidegger, the resort to time (and the future, in particular) to rework the ontological is susceptible to a certain reversal. Time is first introduced for the ontological services it can render, but in order to perform such services, it has to be reshaped and reconstrued in such a way as to shed some of its original characteristics. It cannot, for example, help free us from the grip of a naive ontology if it itself is carrying the same disease. Thus, in discussing Being-toward-death, Heidegger will distinguish expectation from anticipation, where expectation considers death as a real-time event (a funeral on Tuesday) and anticipation considers death as a meaning horizon (or an abyssal loss of meaning). Only the latter opens up an authentic attitude to death. More generally, Heidegger will withdraw from the temporality of programmatic action,[4] he will diagnose our contemporary interpretation of being as that of framing (*Gestell*), and he will propose thinking as an alternative to philosophy in terms of its freedom from calculation, prescribed rules, and schemes. Derrida advances a similar valuation in a number of domains, and it takes an exemplary form in the case of time—both in the way we inherit the past and open the future. A programmed future offers no possibility of responsibility.

The Event of Interruption

I cannot calculate everything, predict and program all that is coming, the future in general, etc. and this limit to calculability or knowledge is also, for a finite being, the condition of praxis, decision, action and responsibility.[5]

Derrida

Whenever something other can arrive, there is a "to come," there is something of a "future-to-come." With . . . determinism . . . there is no future.[6]

Derrida

Anything but Utopian, messianicity mandates that we interrupt the ordinary course of things, time and history here-now; it is inseparable from an affirmation of otherness and justice.[7]

Derrida

Where thinking suddenly stops in a configuration pregnant with tensions [a historical materialist] . . . recognizes the signs of a Messianic cessation of happening, or, put differently, a revolutionary chance in the fight for the oppressed past. He takes cognizance of it in order to blast a specific era out of the homogeneous course of history.[8]

Benjamin

Derrida insists he does not really want even a trace of (or perhaps only a trace of) a religious sense of the messianic in his "messianicity without messianism." If Benjamin is still thinking of special redemptive events, Derrida must take his distance from him, too. And this despite the fact that the values at stake in their discourse overlap strongly; Benjamin speaks of "the fight for an oppressed past" and Derrida of justice and responsibility. The major difference is that for Derrida, messianicity is "a universal structure of experience."

This idea of a universal structure of experience is a concession to a certain transcendental normativity. Indeed, even the word "experience" might be thought to suggest a subject controlling a certain closed engagement with the world. Derrida is undoubtedly thinking of Kant or Husserl yet insisting that we can understand exposure to the other quite as much as a structure of experience, even as it escapes any kind of closure—indeed, precisely because it does. We could understand this move as a synthesis of Levinas's grasp of the way the experience of the other person explodes the structure of intentional immanence, and of Heidegger's sense of being-toward-death as the affirmative acknowledgment of my exposure to an event that will undo me. Neither Levinas nor Heidegger are making ethical claims. Rather, in each case, a certain normativity is being opened up. And yet with them, as with Derrida, there is a real ambivalence that deserves further attention. At one level, any experience that focuses our attention this way rather than that, that reflects a particular interest or commitment, also excludes, shuts out, some other, just as the third interrupts Levinas's face-to-face relation. On this reading, the messianic "to-come" would be the return of the repressed. Or we could say that any experience generates a certain ideality that cannot actually be fulfilled and that "calls for" an event (whether it be a person or not) to complete or fulfill it. Or again, we could say that what an experience will lead to, how it will develop, is always "up in the air," always remains to be determined. Either way, there is an intimate relationship between a certain structural incompleteness of "experience" and the peculiar temporality of the "to-come." These accounts, perhaps a little bare, would allow messianicity to be a universal structure of experience, or perhaps the universal exposedness of any experience. But Derrida goes further when he writes: "Nothing is more 'realistic' or 'immediate' than this messianic apprehension, straining forward toward the event of him who/that which is coming. . . . [T]his experience . . . is at the same time a waiting without expectation . . . but also exposure without horizon."[9] He writes this in the context of dismissing the charge of utopianism. Messianicity is not about some imagined perfection in the future; it is about "desire and anguish" here and now. But surely he only manages to ward off utopianism with urgency and immanence at the price of universality. Can it really be said that every experience "strains toward the event of him who/that which is coming"? Or again, when he writes that "messianicity mandates that we interrupt the ordinary course of things, time and history here-now; it is inseparable from an affirmation of otherness and justice"?[10] It is hard to see this as mandated by a universal structure of experience. This formula certainly distinguishes him from Benjamin, but it does so at the price of overgeneralization. Derrida finds himself in this position precisely because he has tried to dislocate the privilege of the special case (the messianic) by demonstrating that it is universal/transcendental.[11] When this works, it is a powerful move. But it is not actually necessary for his argument. It suffices to say that any experience is, in principle, susceptible to messianic intensification (i.e., open to interruption by the other), that it may present a good opportunity to reaffirm the importance of justice. Or, better, that no experience is a priori ruled out for consideration in this respect. And that this possibility reflects a certain necessary focus and closure of every experience. This would steer us all back to those places at which Derrida reminds us that every intervention is a matter of strategy and risk. Derrida has to, indeed wants to, recognize the distinctive importance of particular events. He imagines a sea change in our treatment of (nonhuman) animals, he focuses very specifically on 9/11 and he is troubled by how to allocate significance to the historical specificity of Christ (whether it could have inaugurated a general messianicity). He imagines, indeed solicits, a New International, a more hospitable Europe, a different university, cities of refuge, a different kind of friendship. Each of these would be an event, or a reinforcing cluster of events, a reinstitution of the very form of an institution. They would change the basic structure of experiences that took place under their aegis. The insistence on there being a variable messianic potency to different experiences does not rest on the power of reflective thought (that is but one avenue) but equally on what Benjamin calls the "constellation" of circumstances. And it would be foolish to downplay the power of those who can point the way—writers, activists, Kafka's "assistants," even philosophers like Derrida himself, who occupy both sides of the Messiah line, prophets in their own right—but also who imaginatively open us up to what might be thought to be the im-possible, prepare us for interruption, for the wholly unexpected.

Derrida is articulating a practice (and a patience) of creative, critical openness predicated on the recognition, simple but profound, that the future never comes "as such." What comes is always another moment with its own mixture of horizons, uncertainties, exposure, and im-possibility, its own manner of taking up / taking on the past and of opening onto "the future." This open, differential, eventuating matrix of thinking and engagement is not a substitute for concrete political strategy, either, but it functions as a source of recursive resistance to both dogmatism and despair.

Given the dangers of thinking "the future" ontologically (Hegel made this clear with his *"Wesen ist was gewesen ist,"* which puts the past in the ontological driver's seat), we might say that serious reflection on time always dooms ontology. It might be thought foolish to ask the question: Does Derrida do justice to calculation here? Everything of value seems to begin once we set aside the program, the law—calculation. As with Heidegger (who insisted that technology was not, for him, the devil), we must not oversimplify Derrida's position.

There is a sense in which he is right to say that ethics (responsibility, justice, friendship, the gift) begins at the point at which calculation breaks down. The demonstration of a (hidden) program, a rule, an economy of exchange seems fatal to our self-understanding here. It may seem that the moral and political imperative should then focus on what escapes these logics. But, taking one more turn round the field, it might be that we need just as much to focus on cases that escape not the program as such but the requirement of purity, of having extirpated any trace of the program, of being pure event. Derrida's strategy, it seems, is to plot the tensions between the rule-governed and our openness to what interrupts us and exceeds expectations. But might there not be a rich vein of complex and aporetic relations to "the future" that are more firmly rooted in the calculable—perhaps in "our" refusal to do the math? Many of the issues raised by the environmental challenges that face us are precisely of this sort. In the way they address questions at once material and transcendental (to the extent that they concern "conditions of possibility"), questions about the boundaries between human and animal, issues of international law, the precarious distinction between killing and letting die, and the possibility of an "impossible" shift in human consciousness and behavior—many of Derrida's concerns come streaming back under a more nuanced umbrella.

Unexpecting the Expected

For what tomorrow will be, no-one knows.[12]

Victor Hugo

I don't think anybody anticipated the breach of the levees. They did anticipate a serious storm. But these levees got breached. And as a result, much of New Orleans is flooded.[13]

George W. Bush

I don't think anybody could have predicted [that someone would try to use] a hijacked airplane as a missile.[14]

Condoleezza Rice

Deputy Secretary Armitage said the United States "miscalculated" the level of terrorism in Iraq, describing the insurgency as more "virulent" than expected.[15]

The situation in Iraq today is the result of a colossal and tragic miscalculation. . . . It serves no one to continue to mislead Congress and the American people on what is possible, the pace at which security can be restored, or the costs.[16]

The idea that we might best prepare for the future by expecting the unexpected or at least anticipating that the unexpected will happen or that we could find a renewal of our ethical and/or political conscience in doing so is not unattractive. It aligns quite well with the social conscience that would see in the future some sort of opportunity for redistributive justice, in which the meek do not exactly inherit the earth but get a fairer share of its resources. But there is a quite different phenomenon that needs equal treatment, which is that of unexpecting the expected, refusing to do the calculations, burying one's head in the sand, and turning a blind eye to the point of culpable negligence. Here the problem is not that the future is unknown but that "we" live in denial. Of course, in many ways, the prevalence of such a structure is not surprising. While coming to terms in an affirmative way with our own mortality is a fine thing, as philosophers from Socrates to Heidegger have proposed, it is much more normal to forget about the whole business. Where the bad news is global and scheduled to occur many years down the track, it soon enters the domain of the discountable and can be kicked into the long grass.

It may seem strange to be talking about calculation at all in reference to the major concern of our age—the environmental crisis and global warming. Is not the weather the classic case of the unpredictable? But next week's weather, it seems, is much more difficult to predict than global trends over the coming decades. We do not have precise measures to tell us when the Gulf Stream will finally stop flowing or whether nonhuman species are becoming extinct at the rate of 50 a day or 150 a day. We do know, however, that unless we change our ways pretty quickly, and dramatically, the planet will cease to sustain life in anything like the way we currently take for granted. When we are "interrupted," only the timing should come as a surprise. The writing is on the wall. The question of calculation, what "we" could have foreseen, what it can be admitted was not foreseen, is political. The future is an essentially contested zone. It is important to remain open to the incalculable, and the unexpected, but it is at least as important to cultivate the institutions and the civic leadership that will take responsibility for the not-so-very-incalculable future. The rhetoric of the incalculable and the unexpected feeds a culture of unaccountable and culpable negligence.

From an unexpected source, I would cite here the work of Peter Singer,[17] a radical utilitarian who is, in many ways, utterly opposed to Derrida's approach. And yet, starting out from a similar skepticism about the difference between killing and letting die, he generates some extraordinary public policy challenges by showing how we might alleviate vast suffering by redirecting resources. These "calculations" (recall Bentham's "Calculemus" [Let us calculate]) are not so much ready-made programs as they are challenges to the moral imagination. The question very often is "Who are 'we'"? Are we Americans or citizens of the world? These are precisely the questions that flow, like melting glaciers, from *The Politics of Friendship*,[18] from *Specters of Marx*, and so forth. Calculation, Singer-style, breaks open the boundaries of the fraternal bond, the self-absorbed community, even our privilege as a species.

The question "Who, we?" has a special relevance to the question of prediction, calculation, and the unexpected. Derrida typically (not always)[19] writes as if this applies to each of us individually. But the political question here has precisely to do with the negotiation of the passage back and forth between the individual and a range of other collectivities, in terms of the horizon of the future (and our ethical stance to what might lie outside that horizon). When public planning so conspicuously and repeatedly miscalculates in a way that ordinary individuals find no problem predicting, one has to question the quality of our representatives' interest in the future.[20] Or their interest in not caring about the future. A devil's bargain has perhaps been struck between those who would extract profit from the future (e.g., by increasing the national debt, payable by our children) and those offering religious salvation in the hereafter.

When Derrida talks about democracy-to-come, it is not, he says, about a future democracy but about an urgency, a present possibility. That is, one might say, a constant actual and possible openness to differences of all sorts, and a willingness to allow them to have a voice. In one sense, of course, this does point toward a possible future democracy, not today but perhaps tomorrow, which would indeed be more open to such possibilities—closer to what Derrida calls the New International. If, as I claim, there is an ethical future in calculation—not perhaps as a program but as a source of both imagination and warning—then we could perhaps come to see that what calls to us does not begin on the other side of calculation but is already intimately involved in it.

Gaming the Public Sphere: Spinning and Framing

But is not global warming, for example, precisely an area in which "scientists disagree"? Scientists do disagree, but not about the trends. The appearance of disagreement is very real and, it seems, the result of very careful planning ("calculation").[21]

People generally do not favor action on a non-alarming situation when arguments seem to be balanced on both sides and there is a clear doubt. . . . Accordingly, means are needed to get balancing information into the stream from sources that the public will find credible. There is no need for a clear-cut "victory." . . . Nurturing public doubts by demonstrating that this is not a clear-cut situation in support of the opponents usually is all that is necessary.[22]

It would be tempting to discount this as a marginal phenomenon were it not for the dramatic gap between what scientists believe about global warming and what the public has come to believe is the case.[23] Here calculation, in the shape of carefully targeted public relations programs, has successfully achieved its goal of sowing doubt about the future and, hence, undermining the public will to make prudent calculations. The recognition that facts are grasped inside the frames within which they are presented is the central idea behind George Lakoff's work,[24] which gives Heidegger's thinking of the *Gestell* a contemporary spin within communication studies.

When Derrida talks about algorithms, programs, and calculation, he is mostly dealing with the limits of the conceptual schemes with which we honestly and fallibly try to negotiate with the world. We need to be aware that we are also up against carefully tried and tested cognitive traps designed to corrupt public discourse and hide from view the need for urgent transformation. The dilemma usually expressed by those who read Lakoff is whether to respond in kind—by counterframing. Derrida would, for sure, ask whether there was a difference in principle between cynical manipulation and our everyday tendencies to unconsciously inhabit oppositional schemes.

Leaning on the Truth

The aide said that guys like me were "in what we call the reality-based community," . . . people who "believe that solutions emerge from your judicious study of discernible reality." . . . "That's not the way the world really works anymore." . . . "We're an empire now, and when we act, we create our own reality. And while you're studying that reality—judiciously, as you will—we'll act again, creating other new realities, which you can study too, and that's how things will sort out. We're history's actors . . . and you, all of you, will be left to just study what we do."[25]

Hannah Arendt often recalls that the liar is a "man of action," and I would even say, par excellence. Between lying and acting, acting in politics, manifesting one's own freedom through action, transforming facts, anticipating the future, there is something like an essential affinity.[26]

Derrida

To lie is the future. . . . To tell the truth is to tell what is or what will have been, and it would be to prefer instead the past.[27]

Derrida

Koyre had no difficulty illustrating this "primacy of the lie" in a totalitarian system (whether proclaimed or not), which more than any other needs a belief in the stable and metaphysically assured opposition between veracity and lie; and we would have no trouble illustrating this today, whether we look near or far away.[28]

Derrida

A report, assembled by Conyers' staff, describes the Bush administration's invasion of Iraq as the perpetration of a crime against the American people. . . . [T]he authors of the report find a conspiracy to commit fraud, the administration talking out of all sides of its lying mouth, secretly planning a frivolous and unnecessary war while at the same time pretending in its public statements that nothing was further from the truth.[29]

Derrida's discussion of the possibility of a "History of the Lie" makes uncomfortable reading for anyone with an investment in a clean distinction between truth and lie. No more so than in the new era of fake news. And time and the future are the main culprits. For to the extent to which performativity is inherent in language, words can create or help create the states of affairs they seem merely to be describing. If we add to this the idea that "men of action" have the further power to bring about states of affairs, then in these cases reality can be made to conform to words, rather than vice versa, "making things true."

It is important to acknowledge that prophetic words can clarify or crystallize a situation in such a way as to open up a new dispensation of the real. We may then argue that all words have some prophetic power, even if it is limited to that of confirming existing frames of reference. What Heidegger thinks of Holderlin's poetry would be true of language generally but writ large. All this is an important corrective to a naive static representational understanding of the relation between words and things. But it would be a mistake to exaggerate the significance of this corrective. In particular, there are two phenomena that cut against an extreme pragmatist view of truth. The first is that the words people use, the promises and predictions they make, do not just have effects and disappear. They are remembered, recorded, written down, and can be returned to and subsequently judged in the cold light of day. The second is that we can attempt to "create our own reality," to reshape the world, and yet fail, exposing the lies as lies rather than as calls to action validated by results. It is not just totalitarian regimes that need to hold on to the distinction between truth and falsity. Democracies, too, need to be able to assess when they are being lied to, even if, as too often happens, these assessments take place retrospectively. To say "the lie is the future" is to tell a half-truth. One could equally well say "the truth is the future," emphasizing the power to make real. But the future can bite back and expose a lie as a lie. Words do not inaugurate realities without resistance.

The Traumatic Event

What is terrible about "September 11," what remains infinite in this wound, is that we do not know what it is and so do not know how to describe, identify, or even name it.[30]

Derrida

We must rethink the temporalization of a traumatism if we want to comprehend in what way "September 11" looks like a major event. . . . The ordeal of the event has as its tragic correlate not what is presently happening or what has happened in the past but in the precursory signs of what threatens to happen. . . . It bears on its body the terrible sign of what might or perhaps will take place, which will be worse than anything that has ever taken place.[31]

Derrida

This weapon is terrifying because it comes from the to-come, from the future, a future so radically to-come that it resists even the grammar of the future anterior. Traumatism is produced by the future, by the to-come, by the threat of the worst to come, rather than by an aggression that is "over and done with."[32]

Derrida

A major event should be so unforeseeable and irruptive that it disturbs even the horizon of the concept or essence on the basis of which we believe we recognize an event as such. . . . [T]he event is first of all that I do not comprehend.[33]

Derrida

Further, the process of transformation, even if it brings revolutionary change, is likely to be a long one, absent some catastrophic and catalyzing event—like a new Pearl Harbor.[34]

Project for the New American Century (PNAC)

Derrida's double inscription of the event—both as a universal messianicity, or openness to the interruption of the other, and as an actual transformation of the condition of such an openness—is not without difficulty. But his account of the traumatic aspects of 9/11 is most convincing. Borrowing perhaps from Blanchot's rendering of disaster (in *Writing the Disaster*), Derrida draws out two different aspects of the traumatism. First, that it disturbs or threatens our capacity to name, or understand, it. We end up naming it with the date on which "it" happened, as if it had overwhelmed language. But second, it appears as a portent of worse to come, transforming the future into something infinitely threatening. It is as if the tension between "affirmation and fear, promise and threat" constitutive of messianicity (at least on one reading) gets resolved diabolically into fear and threat. Openness comes to mean vulnerability. We could add a third aspect, perhaps not strictly traumatic, arguably written on the face of this event from the very beginning—its status as a casus belli, the "very catastrophic and catalyzing event, like a new Pearl Harbor" that had been called for by the PNAC only two years earlier. It was not difficult to see that 9/11 could be drafted into service as the basis for the dramatic curtailment of civil liberties at home and for military aggression abroad. This was wholly predictable and calculable, and, for many, the urgent ethical and political imperative was to try to prevent this future from happening. What this third aspect of cynical exploitation showed was the lethal effect of subjecting a nation's response to a trauma to opportunistic political calculation.

As far as its traumatic effect on the future was concerned, 9/11 was not a unique event. After the Second World War, an even more global and far-reaching future shock was generated by the prospect of nuclear conflagration and was resolved only by the stabilizing logic of mutually assured destruction (MAD), in which no rational actor could be the first to press the button. On 9/11, however, the parameters of the problem were redefined, demonstrating that non-state agents without a "home" to which the threat of retaliation could be directed could be just as dangerous to some particular part of the world without pulling down the whole theater around them. Terrorism seemed unamenable to anything like the MAD logic of containment, and so an even greater source of anxiety.

Our Global Crisis

Consultants to the Pentagon released a report laying out the possible impacts of abrupt climate change on national security. In a worst-case scenario, the study concluded, global warming could make large areas of the world uninhabitable and cause massive food and water shortages, sparking widespread migrations and war.[35]

In *Specters of Marx*, Derrida refers to ten plagues of the New World Order.[36] He called for a New International that would, beyond existing state boundaries, work to address these "plagues."[37] He subsequently willingly assented to naming our global environmental crisis as the eleventh plague.[38] In this book, in many others, and in countless interviews, Derrida clearly demonstrates a grasp of the relevance of calculation, even statistics, for ethical and political intervention. He speaks knowledgeably of "the nearly 50 percent of women who are beaten or fall victim to sometimes murderous abuse (the 60 million disappeared women, the 30 million mutilated women), of the 23 million infected with AIDS (of which 90 percent are in Africa)."[39] What he does not do is draw the ethical and political consequences for our thinking of time and the future from the role this prescribes for calculation—not as setting the limits of action and implementation but providing them with ground, motivation, and priority.

I want to focus here on the global environmental crisis, in particular global warming, for the peculiar engagement with time and the future that it calls for. Let us first begin with the well-worn story of the frog in hot water. It is said that if one puts a frog in a pan of water and heats the water on the stove, the frog will progressively adjust to the rising temperature until it is finally boiled alive. At no time is the incremental temperature difference great enough to alert the frog that something is wrong. Could not the frog's experience have been structured by a messianic openness to interruption that never came? Would it not have been better off calculating and, at a certain point, cutting its losses? What science teaches us here is the significance of changes that may not be visible to the naked eye, and that disaster can stalk us on slippered feet. Having said this, we need to recognize that "science" may never present us with the hard evidence that would make doubt impossible. Not only does the possibility of doubt lie at the heart of science, but there is no "science" of the same order as the traditional discrete natural sciences (biology, chemistry, physics) dealing authoritatively with all the significant processes operating on this particular planet at this point in history. There can be no science of a singularity, because science is in the business of generalization over repeatable situations. But this does not mean—as we have seen—that we should "wait until all the facts are in," as global warming deniers argue, any more than we should wait until our lungs are infested with tumors before we stop smoking. Rather, and here we discover a most distinctive and striking feature of many environmental issues, including global warming, if we wait until we are certain that the process we are concerned about is under way, it will be too late to prevent it from happening. This has led people to embrace the precautionary principle—that where something is likely and is bad news, we should act to prevent it even before we know for sure it will happen. The more sophisticated version substitutes full-blown risk-assessment studies to balance out the level of danger with the probability of occurrence—more calculation. Of course, it could be, and needs to be, argued that we are no more able to guarantee our probabilities than the certainties we began looking for. And when it comes to the end of the world, we do not exactly have "rules" to go by. Is any degree of risk worth taking with the end of the world? Could it really be too expensive, too damaging to our economy, to cut our CO_2 emissions if that were needed to prevent a catastrophic rise in the surface temperature of the earth? We do not get much help here from the to-come, or universal messianicity. It is perhaps just the sort of situation in which we need to go through the undecidable. As Derrida writes, "A decision that didn't go through the ordeal of the undecidable would not be a free decision, it would only be the programmable application or unfolding of a calculable process."[40] Of course, and it is important to say this, we can involve quite as much calculation as we like in our decision-making process. The point is that there is no final algorithm that will enable us to crunch our numbers and values and get a result.

We have suggested that one of the problematic ingredients in the decision mix is the historical singularity of the history of the planet, which makes the issue of global warming more ideographic than nomothetic in character, more a matter of judgment than proof.[41] The absence of repetition makes inductive generalization difficult. But there is another level at which we must be wary of induction. It is often argued that there have been prophets of doom before, and they have been wrong. So why believe them this time? This clearly proves nothing at all. The problem of the Boy Who Cried Wolf was not that the wolf did not come but that when it eventually came no one would believe him. As Bertrand Russell wryly points out, the chickens in the farmyard have inductive evidence of the farmer's benign disposition toward them—until the day he wrings their necks. If doom had happened, we would not be here doubting it. And we have no shortage of lesser dire predictions that have been right on target.

There is no doubt that the difficulty we have in grappling with the prospect of global warming is that most (not all) of our political and personal thought processes concern themselves with more or less correctable or reversible events. Nixon's impeachment was a "triumph of democracy," only if we discount Vietnam. And global warming takes the problem to another level. We have few institutions dedicated to preventing the momentously tragic and irreversible, such as the breakdown of fundamental life processes. Most presuppose that these processes will continue ad infinitum. We are not programmed to think nonlinearly. Even our own deaths make more sense if we can pass things on to the next generation one way or the other. The prospect of global warming, although not a punctual event like 9/11, is equally traumatic, as Derrida describes it, in attacking the fundamental premises on which our capacity to understand or adequately respond are based. A resourceful earth is a condition for everything we are and do, not a familiar focus of concern, at least globally. One more unfamiliar or at least psychologically unsettling feature: global warming is a slow invisible disaster composed of the effects of tiny innocuous acts, most of which are deemed perfectly acceptable by the local culture (driving, heating the house, flying to relatives for Christmas). We have done nothing wrong! And yet, this may be the meaning of Kafka's cryptic remark to Max Brod that there is "plenty of hope, an infinite amount of hope, but not for us."[42] The scale of this "not for us" can be contested. We might conclude that only a quite differently constituted posthuman creature could take forward evolution from this point. This might allow Derrida (or Benjamin or Nietzsche) the opportunity for a more substantial messianic graft in which we would welcome such an Übermensch. More pessimistically, perhaps, we might resign ourselves to the thought that even if we radically curtail the evolutionary process by our toxic behavior, once we humans are gone, the machinery can be expected to trundle on and strike out on new paths. We are engaged in blindly repetitive habits for which the usual penalty is limited: boredom or a wasted life. We are used to the consequences being personal. But in this case, our repetitive habits have cumulative consequences; we are adding grains of sand to scales creaking toward irreversible tipping points.

A New Figure of Europe

This Europe, in its unholy blindness always at the point of cutting its own throat, lies today in the great pincers between Russia on the one side and America on the other. Russia and America are, when viewed metaphysically, both the same: the same hopeless frenzy of unchained technology and of the boundless organization of the average man. When the farthest comer of the globe has been technically conquered and can be economically exploited; . . . then, yes then, there still looms like a specter over all this uproar the question: what for?— where to?—and what then?[43]

Heidegger

There can no longer be a balance of terror, for there is no longer a duel or stand-off between two powerful states (USA,
USSR).[44]

Derrida

I am speaking of a new figure of Europe. . . . Without forsaking
its own memory . . . Europe could make an essential contribution to the future of international law.[45]

Derrida

I would like to hope that there will be, in "Europe," or in a
certain modern tradition of Europe . . . the possibility of another discourse and another politics, a way out of this double
theologico-political program.[46]

Derrida

Derrida's reflections on geopolitical matters did not begin in the
aftermath of 9/11. *Specters of Marx: The State of the Debt, the Work
of Mourning and the New International* (1994) was a scathing in-
dictment of Western liberal triumphalism in the particular shape of
Francis Fukuyama's *The End of History and the Last Man*.[47] In *The
Other Heading: Reflections on Today's Europe* (1992), Derrida had
already linked a democracy-to-come with a critique of the limits of
state sovereignty and the recasting of Europe's privileged position in
forwarding the project of enlightenment as an opportunity, indeed
a responsibility, for hospitality. In this respect, Derrida is not simply joining the ranks of critics of eurocentrism. He is engaging in a
creative *displacement* of the philosophical tradition that would celebrate the privilege of German and/or European world leadership, from
Kant through Fichte, Husserl, and Heidegger, arguing in effect that
the contestation of state sovereignty already under way in the new
Europe could open onto a new sense of global hospitality, a reformu-
lated affirmation of international law, new transnational institutions,
and networks of solidarity and communication. All this he christens
the New International. Europe becomes the site of a possible new

promise. Its privilege would consist of opening "the possibility of another discourse and another politics" in which "it" would no longer be privileged. When he writes, "Without forsaking its own memory . . . Europe could make an essential contribution," he is insisting on the continuing value of distinctively European traditions, even as we must continue to negotiate with the legacy of imperialism, colonialism, and genocide. Paradoxically, the possibility of something radically new may only be opened up by rejecting the temptation to repudiate the past. For such repudiation risks simply confirming and repeating an oppositional frame that offers no escape. And it loses touch with the very values we will need to work with in reimagining the future. This, in a nutshell, was his message in *Specters of Marx*.

What is clear from these kinds of interventions, however, is just how limited, even for Derrida, is the role to be played by the "messianicity without messianism" and "democracy-to-come" in their refined versions, versions that abjure "the future" in favor of a universal immanent structure of experience. We take this up in the following section.

A Rogue State

The most perverse, most violent, most destructive of rogue states would thus be first and foremost the United States.[48]

Derrida

At present the United States faces no global rival. America's grand strategy should aim to preserve and extend this advantageous position as far into the future as possible. . . . And advanced forms of biological warfare that can "target" specific genotypes may transform biological warfare from the realm of terror to a politically useful tool.[49]

Preserving the desirable strategic situation in which the United States now finds itself requires a globally preeminent military capability both today and in the future.[50]

The world order, in its relative and precarious stability, depends largely on the solidity and reliability, on the credit, of
American power. . . . [T]o destabilize this superpower, which
plays at least the "role" of the guardian of the prevailing world
order, is to risk destabilizing the entire world, including the
declared enemies of the United States.[51]

Derrida

What appears to me unacceptable in the "strategy" . . . of the
"bin Laden effect" . . . is, above all, the fact that such actions
and such discourse open onto no future and, in my view, have
no future.[52]

Derrida

The perversity of describing the United States as a rogue state is
perhaps not entirely unconnected with the sense of the French original (rogue = *voyou*) as "cheeky street-arab."[53] The structure of this
back talk is ethical, political, and, let us say, topological. Ethically,
it has the same tu quoque shape as "Let he who is without sin cast
the first stone." Topologically, it echoes the projective blindness
that Derrida diagnoses in Kant's essay "On a Newly Arisen Superior
Tone in Philosophy." He who denounces the apocalyptic may not be
entirely free from its madness. His reading of Foucault has a similar
structure.[54] But its political significance is sobering and must not be
underestimated.

Derrida's deepest diagnosis of 9/11 as an event traumatizing our
sense of the future, replacing the play between hope and anxiety
with a wound that will not heal, is tied in part to his sense of the
United States as a guardian or guarantor of global "order," even to
the point of sustaining the information flow and communication
systems by which this event was filmed in real time and filled public
space. At a certain level, Derrida is acknowledging that the United
States, for all its faults, is holding open the ring within which it
itself is playing out its struggles with other powers. It has what we
might call a materially embodied transcendental role. The question

of whether it deserves this role or whether we could not think of a substantially better way of doing this is, at one level, beside the point. And it is not only holding open a "space" but also time, in particular the future. "The world order, in its relative and precarious stability, depends largely on the solidity and reliability, on the credit, of American power."

This credit (see references in *Specters of Marx* to the "state of the debt") is economic, certainly, but it is also institutional and even military. Even if it acts in this way entirely in its own interests, it is also the guarantor of a global order without which the whole deck of cards might well collapse. Were he forced to make a choice between terrorism and US hegemony, he would choose the latter. As for bin Laden: "Such actions and such discourse open onto no future and, in my view, have no future." And yet . . .

The very title "Project for a New American Century," the neo-conservative blueprint for American hegemonic domination, should make us pause: the idea that the future, that time itself perhaps, could become the property of a particular state. It is as if Fukuyama's vision of the end of history had been exorcised by *Specters of Marx* only to loom back out of the mist rearmed to the teeth. But unless such a state were to achieve an ascendancy only dreamed of by previous civilizations at the point of their imminent collapse (in this case by establishing a "globally preeminent military capability"), this makes no sense at all. It is just such a premise that is affirmed. This is imagined to include "transform[ing] biological warfare from the realm of terror to a politically useful tool."

We can be grateful to the authors of the documents generated by this project for the clarity with which they state their case.[55] They leave no room for doubt. The ideal of universalizable norms has been traduced into a license for the preservation of massive privilege and inequality. Freedom becomes the name of the rule book on a heavily banked playing field. Democracy itself is not immune from being co-opted where there are higher stakes involved. It is in this context that Derrida's recognition of the role of the United States as our actual, on-the-ground guarantor of order becomes so poignant.

For the logic of this quest for absolute domination is surely that it is a genuinely impossible goal that can be expected to provoke seismic instabilities down the road. This is true politically, but it is just as obviously so environmentally. Even if one state could successfully impose its will on all other states on the planet, nature is not so forgiving. And all the forecasts suggest that as an energy-consumption economy, the American way of life "has no future." Returning to the theme of the messianic, the tragedy is perhaps that no state has ever been in a better position to provide decisive leadership in so many areas, that there has never been a greater need for such direction, and that we have never been so let down. It is in this context that the quest for a democracy-to-come and a New International becomes all the more pressing.

A New International

It is necessary for this "new International" to be developed, this engagement . . . between men . . . between living beings (with "animals"!), and then, inseparably, between the living and the dead, and even between the living and those who are not yet born. It does not stop at the borders of the nation state.[56]

Derrida

I am not unaware of the apparently utopic character of the horizon I'm sketching out here, that of an international institution of law and an international court of justice with their own autonomous force. . . . [I]t is faith in the possibility of this impossible and, in truth, undecidable thing from the point of view of knowledge, science and conscience that must govern all our decisions.[57]

Derrida

We know that the Jews were prohibited from investigating the future. This does not imply, however, that for the Jews the future turned into homogeneous, empty time. For every second of time was the strait gate through which the Messiah might enter.[58]

Benjamin

Derrida, as we have seen, insists that we not understand the "to-come" as a real future "down the road" or "around the corner" in a way that itself both repeats and twists free from Kant's invocation of the transcendental (v. the empirical), Heidegger's ontological (v. the ontic), and perhaps Levinas's infinite / the ethical (v. totality / the political). The "to-come," something like the "difference horizon" of every event, a "horizon" that opens rather than frames, can no more be understood in this vulgar way than (my) death can be understood simply as one event among others. But the universality of such a structure, and even our recognition of this fact, is not the same as, and is no substitute for, the creation of new institutions, indeed new forms of institutions or institutional processes.

There are so many ways in which "our" future has been threatened, traumatized, damaged, and compromised, even as we can celebrate more local ways in which the weather still looks good for sailing. The crisis is such that the very idea of "we" is in question. The proponents of the PNAC in effect single out Americans as the "we" and, if the truth be told, Americans who are not poor. The accelerating rate of species extinction poses a huge dilemma—are "we" (we life-forms) all in this together or not? And what are called questions of intergenerational equity expand into broader issues about whether and how we expect our own species—"our" children and grandchildren—to have a future. If we are to "live on," we will indeed need new attitudes, new institutions, for otherwise "the future can only be anticipated in the form of an absolute danger." We should not forget: "Only that historian will have the gift of fanning the spark of hope in the past who is firmly convinced that even the dead will not be safe from the enemy if he wins."[59]

Conclusion

In brief, Derrida insists that we understand the "to-come" not as a real future "down the road" but rather as a universal structure of immanence. However, such a structure is no substitute for the hard work of taking responsibility for what are often entirely predictable and preventable disasters (9/11, the Iraq War, Katrina, global warming). Otherwise, "the future can only be anticipated in the form of an absolute danger." Derrida devotes much attention to proposing, imagining, and hoping for a "future" in which im-possible possibilities are being realized. It is important to steer clear of the utopian black hole, the thought (or shape of desire) that the future would need to bring a future perfection or completion. To avoid the trap set by such a shape of desire, it is not necessary, indeed is necessary not, to reduce the future to a universal structure of immanence. Arguably, the power of Derrida's writings of the past decade had to do with his creative deployment of a productive tension between a certain alterotemporality of immanence and a politics of the impossible. But what is equally disturbing is not our inability to expect the unexpected but the failure of our institutions to prevent the all-too-predictable. Too many of our institutions have conditions of sustainability that are unhealthily insulated from the real world, or indeed coconspirators in the fantasy that we can go on like this. Heidegger's infamous rectoral address could be reread in this light. When the end of the world comes, it will be disheartening to hear again: "Who would have predicted this?"

CHAPTER

10

Beyond Narcissistic Humanism: Or, in the Face of Anthropogenic Climate Change, Is There a Case for Voluntary Human Extinction?

May we live long and die out.
—*The Voluntary Human Extinction Movement*

'Tis not contrary to reason to prefer the destruction of the whole world to the scratching of my finger. 'Tis not contrary to reason for me to chuse my total ruin, to prevent the least uneasiness of an Indian or person unknown to me.
—*David Hume*

Discussions about "Man's Place in Nature" frequently end up in an unhappy place. A standoff erupts between biocentrists such as Schweitzer, who want to treat humans as one kind of living being among others, and anthropocentrists, who evoke some sense of man's transcendence in relation to the natural world. It is not helpful that this tricky juncture is peppered with aporias and confusions. I try here to chart a clear course through them.

By way of orientation, the more general reflective and methodological questions lurking in the background are:

1. In what sense does the recognition of our distinctive cognitive capacities require a privileging of "the human," what has been

202

called epistemic anthropocentrism more generally, and how should we parse that experience? Is it legitimate, if it is even avoidable, that we be judge and jury in this case?

2. When we speak of distinctive human capacities (reason, intelligence, consciousness), do they deserve acknowledgment whether or not they are realized in practice? (Are rational beings that do not act rationally really rational beings?) Is it sufficient that these capacities be *available*, however undeveloped? And can we claim the privileges of greater insight without shouldering the responsibilities they entail? Finally, could human beings come to recognize that the values that most justify their privileged status would require them to bring the *Homo sapien* experiment to an end?

These questions are clearly both metaphysical and political, and this last set of questions will become our focus.

From a biocentric perspective, humans are living beings who have developed special capacities to better realize desires, goals, and drives rooted, one way or another, in our animal existence. This need not involve a simplistic reductionism. Even if "in the last analysis" reproduction, nourishment, shelter, and survival are the most powerful forces, the realization of these fundamental goals in the context of human society gives rise to secondary goals—respect, power, wealth, knowledge, and freedom, which take on a life of their own, even if their primary role, so to speak, was to protect or enhance, or more surely secure, these basic ends. It would be a matter of critical concern if and when this second nature were to develop in such a way as to threaten the satisfaction of basic biological needs. Where such worries become serious, biocentrism inevitably expands its horizons to take in the interactions and interconnectedness of different life-forms. Whatever value we may wish to attribute to the lives of individual animals or particular species, including our own, a condition of such values being realized is that background ecological requirements are met. This is not in itself a value judgment but a conceptual consequence of the fact that individual organisms are essentially interdependent, both with respect to other species, to

other members of their own species, and to the physical world in
which they find themselves. (The dependency really does work both
ways. The lizard may rely on the rock, but if the rock is limestone, it
is itself the calcified remains of countless sea creatures. A diamond
is the highly compressed legacy of long buried vegetation.)

Essential interdependency is a truth about life that precedes any
agreement on the precise character of that interdependence. It does
not deny, for example, that there is redundancy in nature. Indeed, it
probably requires the opposite—that not every item matters, even
though we may not be in a position to determine which, if any,
items are critical. But (inter)dependency rightly insists that forms
of life have sustaining conditions, whether or not we can agree on
what they are. Relationality, in other words, is constitutive, not a
secondary phenomenon and not just with respect to the intraspecies
dependency of being born to a mother, but the interspecies dependencies of struggle and cooperation, as well as the ways each species depends on others for nourishment and for various ecological
services. This interdependency is current and ongoing in the sense
that all life is essentially engaged in daily exchange with living and
nonliving matter. These exchanges occur at many levels—symbolic
exchange, collaboration in reshaping matter (termites building a
mound), ingestion (breathing, eating), and so on. But this dependency is importantly historical in a deeper sense. When I look at
my keratinous fingernails and scratch my head, I cannot but register
a profound evolutionary bond to human and prehuman ancestors.
Evolutionary heritage speaks of deep ontological dependency on beings and circumstances long dead, even if many co-descendants live
on. Is it important, and if so why, to acknowledge such dependency?
Is acknowledgment part of completing the inheritance? Can there be
adequate/inadequate forms of acknowledgment? On my behalf? On
our behalf?[1] And how should we connect this question of heritage
with that other abyssal dimension, the only half-thinkable future,
where discontinuities and singularities that would sink any capacity for rational projective consideration may await us?

Many discussions in this area lead to one party accusing the other of anthropocentrism—treating or thinking about animal life and the natural world more generally from a human point of view. But what is anthropocentrism? It might be said, for example, that when we speak of creativity, insight, freedom, and so on as values, indeed accomplishments, ones that might begin to justify, even redeem, some of the negative aspects of our impact on the planet, these are surely virtues that *we* value, but it is not clear why the planet or its other inhabitants should value them.

There seem to me to be four ways of defending anthropocentrism.

1. Unabashed anthropocentrism. We are humans; hence, we are entitled to promote and project our own standpoint and interests. (Compare Richard Rorty's *we* are (typically) WASPS, White Anglo-Saxon Protestants). Every other species surely does the same. Enlightened Martians or Venusians would, too.

2. Why not mammalocentrism? The perspective in question is indeed one *we* are articulating, but it does not necessarily reflect the narrowly human standpoint. It might be mammalian. Of course, going in the other direction, it might be that of white, male citizens of developing Western nations.

3. Man transcends his own interests. The perspective in question has a value and significance that transcends our species self-interest. Even if by virtue of tools developed for other purposes we humans do indeed have an objectively superior perspective, we are now the vehicle for its realization.

4. Anthropocentrism is logically unavoidable. Anthropocentrism of some sort is logically unavoidable if the perspective announced is the product of human reflection.

In his essay "The Other Heading,"[2] Derrida argues that one could imagine endorsing a privilege to Europe as the leading edge of a certain historical "progress" if it were to offer itself as an extended "city of refuge," if it offered hospitality to all who needed it. I want to suggest something parallel with regard to anthropocentrism—that its privilege depends on just how it understands "man." I will explore this line of thought by commenting critically on these four defenses of anthropocentrism.

Unabashed anthropocentrism. Anthropocentrism, on this formulation, aims at honesty, at a certain confessional modesty: "Let's not kid ourselves, we *are* human, and it's not inappropriate that our judgments reflect that." Moreover, attempts to take account of the other's perspective risk an appropriation of the other that is blind to this natural prejudice. Is it not better, more honest, to say, "This is where I stand" rather than insisting on second-guessing how *we* stand?

Politically, this position does have some merit. Coupled with democratic empowerment of the other, the others, it gives a premium to respect for others, in their own voice. I do not need to take responsibility for producing the whole picture. But I can more modestly make an honest contribution to it.

However, I have two doubts about such a position.

The first is that its plausibility presupposes a political arena in which a multiplicity of voices can fairly participate. And this arena cannot merely be notional. We can perhaps imagine, with Latour, a parliament of things, but we cannot suppose we are being just simply by acting in a way that anticipates its actuality.

The second objection applies to Rorty's original as well as to the analogy we are drawing with it. It is a mistake to suppose that we can at best attest to a certain solidarity with our own kind. First, because it is genuinely contestable who "our kind" are. Solidarity is elastic in scope—our gender, our race, our nation, our class, our species are all candidates. On what grounds could one be definitively chosen over another? And, just as problematically, it is impossible to determine in some neutral fashion what shape that solidarity should take. If one were to identify the human with its scientific (or religious) achievements, it would take a very different—probably elitist—shape than would a concern with global social justice. A focus on human self-interest just begs the question—politically and in other ways—long before one convicts it of parochialism.

Why not mammalocentrism? The second defense of anthropocentrism is to argue that it is misunderstood as *anthropo*centrism. Promoting community may reflect our status as animals living together. Promoting family values may reflect a certain mammalocentrism rather than anthropocentrism. Valuing complex communicative behavior is something we share with apes, dolphins, and dogs. Reason, intelligence, even consciousness may not be uniquely human. Dolphins can recognize themselves in a mirror. Privileging life over inert matter may well reflect our status as living beings.

With this last example, an important clarification can be made—one that will usher in the next defense of anthropocentrism. If it is said that it is *only because we are living beings* that we privilege life over, say, rocks, or clouds, or a complex star system, it sounds as if the charge is that of irrational favoritism. But surely there is a more intimate connection between what we are (living beings) and the preferences we have, the values we promote. The more intimate connection is that it is only with living beings and their teleological orientation that value arises at all, whether it be the valuing activity itself or the acquisition of value by whatever is the object of valuation. If living and valuing are co-originary, would it be blind narcissism to give some sort of privilege, or at least evaluative primacy, to living beings? Or would it be the appropriate acknowledgment of the source of all value? Setting aside for a moment the possibility of spiritual but nonliving beings (angels, gods, galaxies, etc.), this would be saying that the universe itself lacks value in the absence of the valuation that begins with life.[3]

This, then, is a somewhat strange defense of anthropocentrism, arguing that some at least of what we take to be mere human projections may not be human at all but shared by other species and/or life itself, and that it may be wrong to think of some of these basic values as projections at all. They may simply be articulating the scope and origin of value as such.

Man transcends his own interests. The third defense of anthropocentrism would be to argue that humans do indeed have a privileged perspective, but that that privilege, roughly speaking, is intimately tied to transcendence of any shallow or narrow human self-interest. This, I believe, is the logic of Heidegger's position. And, for that matter, Kant's and Hegel's. For Heidegger, man, qua Dasein, is the site for the first appearance of freedom, truth, and ontological self-interrogation. The privilege of the human is not to privilege *the human* but to have a certain access to being. On this model, birds are justified in privileging flying because of what flying truly enables them to see. The shape of the usual worry about that sort of claim would be that valuing "seeing from above," what Merleau-Ponty calls the *pensée du survol*, is, in fact, a special mode of appreciation, one that birds are good at but that also has its limits and drawbacks, and to ignore those would be to close off a necessary line of questioning. This, indeed, is the tenor of Levinas's critique of Heidegger—that even if Heidegger escapes a narrow humanism, he remains trapped in an ontological neutrality blind to our ethical exposure.[4]

Anthropocentrism is logically unavoidable. Finally, there is the defense of anthropocentrism that could be called logical in the sense that all claims *we* make necessarily reflect our human cognitive and expressive capacities. But anthropogenic need not imply anthropocentric. Even the most biocentric, anthropofugal view is anthropogenic in the sense of coming from man, drawing on our human capacities. Does anything substantive follow from that?

Drawing together the different threads of these justifications for a certain anthropocentrism, a distinction is taking shape between a vulgar (or shallow) and enlightened anthropocentrism.

Vulgar anthropocentrism would either model the value of other creatures on what we value about man (e.g., giving us a scale based on intelligence or reason), or it would value the earth as a whole and its systems and other inhabitants in light of their contribution to our human self-interest, however myopically grasped.

What would an enlightened anthropocentrism look like? Such an anthropocentrism would draw upon what may well be uniquely human attributes to construct an account of, or a conversation about, man's place in nature or options for a sustainable future. As proof that this would no longer be vulgar anthropocentrism, it seems to me that it must be *possible* for enlightened anthropocentrism to conclude, perhaps sadly, that despite being uniquely gifted analytically and imaginatively in being able to understand the situation, there is a dark side to this and/or allied capacities, which renders our human presence toxic to the planet. And it must be *possible* that such an analysis would recommend the termination of the human project, its modification, or its posthuman redirection (e.g., by gene therapy).[5]

One might respond that this is not a remotely possible outcome. Surely, the value of a species selfless enough to make such a recommendation is incontestable! Indeed. But if this genuine strength is linked, contingently or otherwise to an uncontrollable and toxic disposition, then this value is not unconditional. Is an enlightened anthropocentrism that could contemplate such an outcome truly anthropocentrism? I believe so.

If, for example, human beings prided themselves on their capacity for visions of harmony, community, or "the whole," then the kind of logic deployed in war or revolution—"Give me liberty, or give me death"—would apply equally well here.[6] Human existence is the site at which values higher than mere life or survival are born—for Heidegger, the site at which a certain openness to Being happens.

Such a thought could cut two ways. As things stand now, a certain interpretation of "freedom," one that centrally promotes the unrestrained reshaping of nature, might be thought responsible for the slow death of the planet. But might not the pursuit of something glorious be worth it even if it meant finally going down in flames? Is not greatness attended by great risks?

Conversely, if we came to view our sociogenetic constitution as flawed, lethal to the planet, including ourselves,[7] could we not conclude that it would be better for all concerned, *and for the values we most care about,* if we gracefully bowed out? Identification with supremely human values would make this position anthropocentric.

Adumbrating this possibility—of winding down the human experiment—is not even remotely to begin to recommend it. It is simply to mark the possibility of an understanding of "man" as the site of a certain difference, indeed struggle. I mentioned Heidegger's version of this, and it would be instructive to think through Heidegger's being-toward-death in this light. Equally, we might recall Nietzsche's claim that man is something to be overcome. How, he asks, can we seriously endorse an evolutionary account of life and think that we are the final stage? The capacity of the sage or prophet to grasp this possibility of further transformation makes our failure to accomplish it all the more tantalizing. (See Zarathustra's disappointment in Nietzsche's *Thus Spoke Zarathustra*.)

I referred to "something like" Heidegger's account, and now I have alluded to Nietzsche's. Strangely, one could treat Heidegger's account as itself a development of Nietzsche's, moving past the seductions of willfulness toward a genuine receptivity in the shape of *Gelassenheit*.[8]

The idea that deploying our rational powers might lead us to conclude that the earth would be better off without humans presupposes that our planetary presence continues to have its current toxic impact.

There are many reasons to suppose that this will indeed be the case, reasons that could be summed up by the diagnosis that human life has been captured by a powerful and dangerous logic, that of capital, which, even before we lament its consequences for human exploitation of other humans, rests on the extraction of value from natural capital in a way not too different from a Ponzi scheme. Moreover, while this logic operates globally, it is subject to minimal global vision or restraint. If anything, it must subvert any attempts to restrain it. Concern for the quarterly bottom line make it increasingly blind to considerations of sustainability.[9] Plugging in high discount rates makes planning for an increasingly unpredictable future less and less economically justifiable. The unsustainability argument claims not only that we cannot rely (as people such as Julian Simon believe) on markets to infinitely extend natural resources but that sinks like the oceans and the atmosphere are simply irreplaceable in the roles they play. Moreover, this diagnosis is increasingly compelling even to mainstream thinkers. As an example, I would cite James Gustave Speth's *The Bridge at the Edge of the World*, an extremely sober analysis "of systemic failures of the capitalism that we have today."[10] Speth's hope is that with changing corporate incentives, using markets for environmental restoration, that things can be turned around, that capitalism having created the problem can deliver the solution. But this rests on the possibility of social control of a system for which resisting such control may be an imperative, and just another cost of doing business.[11]

The diagnosis of the difficulty of a change of direction rests on taking corporate interests seriously. But equally, the problem lies quite as much with our own habit structures, patterns of motivation, psychosocial investments, and individualistic subject formation, which direct us in so many ways. While they can at times pull in different directions, there is surely a certain deep alignment between tendencies to narcissism at both the individual and species levels.

Habits are adaptations to both the physical and social world, and in their insistence they resemble the effects of trauma, as if being-in-the-world at all were already a traumatic condition. It is from such a frame of reference that I want in this next section to briefly reprise (see Chapters 5 and 6) a number of symptomatic philosophical analyses in which something of a countertraumatic reversal of this fundamental narcissism is being adumbrated.

This structure of switching or reversal is found by Levinas in my coming up against the limit of my own intentional orientation, its interruption by the face (or appeal, or call) of the Other, itself putting a strange reverse spin on Sartre's account of the effect of the look. It is found in Rilke's description of the experience of being looked at by a tree; in the self-disgust Lawrence feels after throwing a stick at a snake; in Derrida's reflections on being looked at, naked, by his cat; and in the various meditations—from Plato, to Hegel, to Nietzsche and Bataille—on the significance of the sun.[12] As we have seen, there are many more sites for such reversals: the enemy, the friend, the master, the ghost, and so on. In each case, the category of the other can emerge or subside without reason, and by such transformations the world is transfigured.

The point of referencing such examples, worked and unworked, is to demonstrate that the phenomenon of renewal and transformation in the self/other relationship is far from being restricted to the charismatically ethical cases of the other in need—widow, orphan, stranger. And the reason for this provides an ontological ground even for those privileged ethical cases, undermining Levinas's claim that ethics is first philosophy.[13]

I understand these instances of reversal as glimpses from within a certain habit structure, of the background truth that these habits occlude. The challenge they present us with is this: do these experiences actually open up new ways of being, or are they just the seductively alluring shadows cast by any everyday practice that could no more constitute an alternative practice than would driving on whichever side of the street took our fancy? My stress on what I am calling their ontological rather than ethical significance points in the direction of saying that the kinds of dependencies they reveal are not things we can optionally acknowledge. We can only set them aside for so long. I agree that there is always the Nietzschean question—how much truth can we live with? But the question runs the other way, too: how long can we live in denial?

In the past few decades, something dramatic has occurred, something that could not perhaps have been anticipated. Human beings have always dreamed of better worlds, of the city on the hill, of some sort of redemption from the wheel of poverty and despair. With some exceptions, however, carrying on in an unreconstructed way was always the fallback position. With global warming and the accelerating destruction of biodiversity, it is becoming increasingly plausible that the status quo ante is disappearing as any kind of option. There has been a shift from thinking about the amelioration of the human condition toward sustainable survival. The reason for this, as I see it, is that we are clearer than ever about our ontological condition—more about that in a moment—even as the collective consequences of our practices belie these conditions.

Human dependency, by which I mean the dependency of the human on the nonhuman, has been very clear in many traditional cultures, whether based on hunting or farming. It is deeply embedded in much myth, not to mention cave-painting. That interdependence was at times confined to specific wild animals (e.g., jaguar, peccary, buffalo, wolf, bear) or domesticated animals (e.g., cow, sheep, pig, cat, and dog), while at other times there was a grasp of a deeper and more widespread connectedness with the natural world, often marked by ritual practices. It may be a cliché, but it is hard not to see urbanization and industrialization as going hand in hand with a breakdown in that understanding, such that it is often refused and/or misrecognized. More to the point, whatever people may say they believe, their practices belie their beliefs. We come to recognize dependency only in the form of trying to subject the natural world to human control. Metaphysically, we might think of dualism as a refuge from the uncertainties of that dependency. It is hard not to think "misrecognition" when Wittgenstein writes: "Man is a dependent being. That on which we depend, we may call God." There is perhaps a double misrecognition here—both in the characterization of the other partner in the relation and in the nature of that dependency.

The two errors are brought together in the atavistic thought that we can effectively deal with the limits of our own control by ritual propitiation of the gods.

Individually and collectively, we are indeed dependent on background conditions that we cannot wholly control. We are vulnerable to storms, flooding, disease, crop failure, and extreme temperatures. But alongside and often superseding religious responses to this condition, we have developed technologies of empowerment and security, which even as they often achieve greater local control (protection from the elements, greater terrestrial mobility) have far-reaching consequences that we cannot control (climate change) and that threaten much greater destruction. They threaten destruction because they influence precisely the ecological systems on which we depend and which cannot be bought off by ritual sacrifice but only addressed by a global transformation of our practices. Our dependency is very real, and for most of the history of humanity we have been *locally* vulnerable. What is new is that local vulnerability is morphing into something of global proportion. The question is: Can we respond adequately to this (new) grasp of our dependency on background conditions (and relational interdependency with other beings)? What is at stake in this question is, strangely enough, more than our survival. One eccentric way of putting it would be that what is at stake is *our right to* survive.

Those who point to our distinctive capacities as humans will typically refer to our capacity for rational thought, reflection, or self-consciousness. It is tempting to suppose that privileging these capacities merely reflects our strengths in this area. As I suggested before, birds might argue for the distinctive value of unaided flight, and fish for swimming. Is there a non-question-begging way of reasserting our privilege? As much as I fear falling into a higher-order myopia of species self-promotion, and despite a hive of doubts, I cannot get away from the thought that humans are uniquely gifted with access to what Heidegger calls the "as such." This does not just mean that we see the rock as a rock while the basking lizard does not (I actually contest that claim) but that we can and do ask questions such as: Why is there anything rather than nothing? What is it to be human, or to be alive? These are not the special property of an intellectual elite with time on their hands but of humans everywhere.

But if it was ever sufficient that we merely *pose* these questions, it is no longer so. For it is the manner in which we pursue them that affects not just our survival and flourishing but that of all our terrestrial fellow travelers. The manner in which we address the question of what it is to be human will determine what it will have been for us to be human. I would like to say—and this touches on the idea of our right to survive (parallel to the move Derrida makes in *The Other Heading*)—that our privilege, as humans, rests not merely on our capacity, however distributed, for wonder and metaphysical insight, glorious though that is, but in our ability to enact the implications of those insights.[14] I am not sure how to demonstrate this, but I am tempted to suggest that wonder is the momentary glimpse of a complexity in depth that needs careful articulation to be brought out, whether its object be existence or the moon or (with Irigaray) the beloved.[15] And that this articulation has consequences.

There are all kinds of reasons (bad reasons) for promoting metaphysical dualism or theories of substance that would downplay what I would call constitutive relationality. If what is constitutive is nonetheless unreliable, it is tempting to deny the relation or transform it into something with an unchanging counterpart, like God. God is the father who will never let you down. But this suggests that there is still some way to go in completing the enlightenment project marked by Feuerbach, Marx, Nietzsche, and Freud, in which we recover from their mythical disguises the true state of our dependencies, the better to affirm and address them.

In a recent proposal for a new book series on animality, the editors wrote: "[W]e genuinely grasp our humanity only through a reflection on our relationship to animality, and by seeking to distance ourselves from animals we render unattainable the goal of ethics, understood in the Heideggerean sense, of finding our place within the larger cosmic scheme of things."[16] This formulation correctly attests to the deconstructive power of constitutive relationality. A relation conceived as "external" is dramatically reconfigured when it is recognized that the terms of the relation ("man," "animal") are intimately caught up in the relation itself. This account needs supplementing by reference both to "animals" and to the matrix of global ecosystems more generally, but it does highlight what is at stake in reflection on anthropocentrism. Moreover, it captures, in an important way, the place of the ethical in the scheme we are proposing. When we speak of man's constitutive relationality and interdependence, it is not to recommend a more just dispensation, a kinder path of greater consideration (though doubtless that would be true). Constitutive relationality is an ontological truth, not an ethical prescription, one that we dismiss at our peril.

The thrust of my argument is this: if, indeed, human beings do have a privileged position in this corner of the cosmos, it is in virtue of a capacity that needs realizing to be given full credence. One objection to using reason (or freedom) as our distinctive and privileged attribute is that it is sometimes blatantly self-serving, but at other times such "virtues" have a mixed track record.[17]

Nonetheless, I suggest that there is not just room for—but need for—an enlightened anthropocentrism, one with three important features:

1. It would take seriously man's heteronomy, our interdependence with other living beings, and our constitutive relationality. We and they survive and flourish only if we manage to sustain the life-support ecosystems that sustain us.

2. This means articulating and implementing what we might call a critical humanism, one for which man is the site of a question, even if the most urgent shape of that question is how to shape a sustainable future in the more-than-human world.

3. Relationality and interdependence are profoundly different kinds of connections with the world from those of projective mirroring. Narcissistic anthropocentrism values the world insofar as it is like us, or suits us; enlightened anthropocentrism, founded on interdependence, mandates attentiveness to the very different ways and shapes of other creatures and forms of life. Enlightened anthropocentrism does not give us a list of creatures we need to cultivate and those we could do without. We may not be able entirely to avoid implementing such preferences—for example, trying to wipe out dengue fever—but it would not be inappropriate to draw on the religious proscription of hubris, connecting it with a broader cautionary principle, which would take the form of a generalized respect for difference, reflecting both our abandonment of the demand for other

beings to be like us and the recognition of the limits of our knowledge. We do not, perhaps cannot, know which parts of the whole we can do without.[18] An enlightened or critical anthropocentrism would pride itself on its refusal to project onto the world either a naive model of man or a reductive understanding of nature. The power to refrain from that is, most likely, distinctly human. I do not deny that we *might* be able to transition to a sustainable world, one in which there were many fewer nonhuman creatures, where we had largely "conquered" disease, and where we had pretty much gotten nature under control in a way that suited us. I am, however, deeply suspicious of this supposition. The fact that we can just about control what happens in a petri dish does not license scaling up to the planet. It is too tempting to underestimate complexity when pursuing an outcome that depends on it not mattering. If we are wrong about that brave new world, there will be no turning back. True enlightenment comes from recognizing the limits of the Enlightenment.

A final comment: it might be thought that I am giving too much weight to implausible possibilities—that humans could come to will the end of their own species. It is not a crazy thought if we come to believe first that what we call our "reason" all too easily metastasizes into narrow calculations of self-interest blind to the collective consequences of everyone acting in this way; or second, that despite the distinctive virtues we possess, we are also wedded to greed, overconsumption, an unsustainable exploitation of natural resources; or lastly, that despite something like reason being a widely distributed personal and collective achievement, there is no longer (if ever there was) an effective general deployment of these powers with respect to the environmental crisis that faces us. This gap between promise and performance has many different sources—an inadequate model of man's place in nature and a failure to realize collectively and practically what the wisest already know. Specifically, lessons that served us well when dealing with local planning and control cannot be relied on if they presuppose externalizing ecological costs to an "outside" that is no longer there, sinks that have been exhausted.

I want to conclude with another—this time more positive—version of this sober assessment. In discussing a Rorty-esque defense of anthropocentrism, I argue that even if one were to value unapologetic solidarity, nothing definitively tells us whether to express solidarity with our species, or our race, or our gender. What this means is that if solidarity with one's own were to triumph over other's considerations (e.g., celebration of difference), and if "the human" came to be seen as an unsustainable project in its current form, it would be open to us featherless bipeds to affirm our solidarity with the life-stream to which we surely belong.[19] We would not need to resurrect or rework religious eschatology, supposing perhaps that after the human something "higher" might emerge. It would, I think, be sufficient to will the creative potential of life, seeing ourselves as one experiment among a myriad of other possibilities. This capacity for stepping back, for imagining otherwise, is surely quintessentially human!

Acknowledgments

This book has been long in the making. It has benefited from invitations from friends and colleagues to present and contribute papers as well as from opportunities presented by numerous conference occasions, especially SPEP, IAEP, and the Collegium Phaenomenologicum in Italy.

Just as importantly, this book would not have happened without conversations over the years with numerous fellow travelers and co-inspirators: Barbara Muraca, Richard Kearney, John Sallis, Dennis Schmidt, Nancy Tuana, Charles Scott, Ted Toadvine, Irene Klaver, Beth Conklin, Keith Ansell-Pearson, Clare Carlisle, Jacques Derrida, Peter Steeves, Michael Naas, Al Lingis, Brian Schroeder, Matthias Fritsch, Phil Lynes, Ed Casey, Jessica Polish, Aaron Simmons, Catherine Keller, Brian Schroeder, Leopard Zeppard, Kelly Oliver, Michael Bess, Cary Wolfe, Jonathan Reé, Jason Wirth, Dimitris Vardoulakis, Chris Drury, Barbara Hahn, John Llewelyn, Brian Treanor, Steve Vogel, Wendell Berry, and Mick Smith. My students at Vanderbilt have provided lively feedback on many of the ideas presented here. Some of these chapters are revised and reworked versions of papers published elsewhere.

I am especially grateful to the late Helen Tartar for first commissioning the book, and to both Tom Lay and Eric Newman for shepherding it so carefully into the world. I would also like to thank

Amanda Boetzkes and a second anonymous reader for Fordham for their extremely helpful critical suggestions for revision, and Jeff Shenton for his insightful comments on an early draft. My special gratitude goes to Nathan Wirth for permission to use his wonderful cover photograph. Thanks, too, to Arc Indexing for preparing the index.

Notes

Introduction: Reinhabiting the Earth

1. Quoted by Martin Heidegger in "Language in the Poem," in *On the Way to Language* (New York: Harper and Row, 1971).

2. Ludwig Wittgenstein, *Philosophical Investigations* (Oxford: Wiley-Blackwell, 2009).

3. US News and World Report. "Revenge of the Weeds," Common Dreams, July 15, 2010, https://www.commondreams.org/views/2010/07/15/revenge-weeds.

4. Gilles Deleuze, *Spinoza: Practical Philosophy* (New York: City Lights, 2001).

5. Martin Heidegger, *Contributions to Philosophy (of the Event)* (Bloomington: Indiana University Press, 2012); "Letter on Humanism," in *Martin Heidegger: Basic Writings* (New York: Harper and Row, 1993).

6. See Felix Ravaisson, *Of Habit* (London: Continuum, 2008). He makes it clear, as we will stress later, that habit has a necessary, constitutive dimension as well as being the enemy of reflection. This is also the case with David Hume and his "custom and habit."

7. See Herman Daly, *Beyond Growth: The Economics of Sustainable Development* (Boston: Beacon, 1997).

8. George Lakoff, *Moral Politics: How Liberals and Conservatives Think* (Chicago: Chicago University Press, 2002).

9. Gilles Deleuze and Felix Guattari, *Anti-Oedipus: Capitalism and Schizophrenia* (Harmondsworth: Penguin, 2009).

10. Compare Descartes's claim that God would never allow wholesale deception.

11. See Pierre Bourdieu, *The Logic of Practice* (Oxford: Polity Press, 1990).

12. Heidegger, *Contributions to Philosophy (of the Event)*.

13. Preface to Friedrich Nietzsche, *Gay Science* (New York: Vintage, 1974).

14. Martin Heidegger, "The Question Concerning Technology," in *The Question Concerning Technology, and Other Essays* (New York: Harper and Row, 1977).

15. From "A Winter Evening," quoted in Martin Heidegger in "Language," in *Poetry, Language, Thought* (New York: Harper and Row, 1971), 201.

16. The fourfold figures prominently in Heidegger's "The Thing," in *Poetry, Language, Thought*.

17. Martin Heidegger, *Being and Time* (New York: Harper and Row, 1962).

18. Karl Marx, "Theses on Feuerbach," in *The German Ideology, including Theses on Feuerbach* (New York: Prometheus, 1993).

1. On the Way to Econstruction

1. An earlier, more theologically oriented version of this paper appeared in Laurel Kearns and Catherine Keller, eds., *Ecospirit: Religions and Philosophies of the Earth* (New York: Fordham University Press, 2007).

2. The considerable resources that deconstruction offers environmental thinking are on display in Matthias Fritsch, Philippe Lynes, and David Wood, eds., *Eco-Deconstruction: Derrida and Environmental Philosophy* (New York: Fordham University Press, 2018).

3. Jacques Derrida, *Limited Inc.* (Evanston, IL: Northwestern University Press, 1988).

4. By a logic that strangely reverses this structure, there is a certain convergence between what Derrida is saying about Rousseau and what Rousseau himself already says, as if Derrida's "supplement" to Rousseau ends up showing that Rousseau knew it all along and did not need Derrida to bring it out.

5. I hint at this in my "Thinking with Cats," in *Animal Philosophy*, ed. Peter Atterton and Matthew Calarco (London: Continuum Press, 2004).

6. See my "Comment ne pas manger: Deconstruction and Humanism," in *Animal Others*, ed. Peter Steeves (Albany: State University of New York Press, 1999), and Derrida's "The Animal That Therefore I Am (More to Follow)," trans. David Wills, *Critical Inquiry* 28 (Winter 2002): 369–418.

7. It seems, too, that many Christian fundamentalists are allergic to this term. It is not clear whether their difficulty comes from its historical connection with liberal politics that they reject or from their resistance to the need to refocus relations of obligation and dependency away from God and toward the natural world. This scene is increasingly complicated by the emergence within evangelical groups of a commitment to "creation care." The question of "humanism" (and antihumanism; see Jean-Luc Ferry's *The New Ecological Order*) is a vexed one. There are those who seem to have proposed that the earth was better off, and would be better off again, with 10 percent of the current human population. This view has been attributed to poet Gary Snyder, who certainly did argue for "[a] healthy and spare population of all races, much less in number than today" ("Four Changes," *Mother Earth News* 1, no. 1 [1970]) and wrote: "The whole population issue is fraught with contradictions, but the fact stands that by standards of planetary biological welfare there are already too many human beings" (*A Place in Space* [Berkeley, CA: Counterpoint, 2008]). But it is a stretch to conclude from this that Snyder is advocating the forced decimation of the human species. What is certainly worthy of discussion is whether the world is a better place, other things being equal, if there are more (happy enough) people in it. Does it really make sense to add up happiness?

8. Jacques Derrida, *Specters of Marx: The State of the Debt, the Work of Mourning, and the New International*, trans. Peggy Kamuf (New York: Routledge, 1994), 85.

9. These "plagues" are: (1) unemployment (in a new sense, social inactivity), (2) exclusion of the stateless, (3) economic war, (4) contradictions of the free market, (5) burden of national debt, (6) arms industry, (7) nuclear proliferation, (8) interethnic wars, (9) mafia and drug cartels (phantom states), and (10) limits of both the concept and practice of international law; emphasis original.

10. For reference, the seven biblical plagues were: (1) land: ugly and painful sores broke out on the people who had the mark of the beast and worshiped his image (16:2); (2) sea: it turned into blood like that of a dead man, and every living thing in the sea died (16:3); (3) rivers and springs: they became blood (16:4); (4) sun: The sun was given power to scorch

people with fire. They were seared by the intense heat, and they cursed the name of God, who had control over these plagues, but they refused to repent and glorify him (16:8); (5) throne of the beast: His kingdom was plunged into darkness. Men gnawed their tongues in agony and cursed the God of heaven because of their pains and their sores, but they refused to repent of what they had done (16:10); (6) great river Euphrates: its water was dried up to prepare the way for the kings from the East (16:12); (7) air: The great city split into three parts, and the cities of the nations collapsed. God remembered Babylon the Great and gave her the cup filled with the wine of the fury of his wrath. Every island fled away and the mountains could not be found. From the sky huge hailstones of about a hundred pounds each fell upon men. And they cursed God on account of the plague of hail, because the plague was so terrible (16:17).

11. Derrida himself quickly and publicly accepted this suggestion (in the audience) when I first made it in my presentation "Globalization and Freedom," *Returns of Marx* conference, Paris, March 2003. This appears as Chapter 10 of my *The Step Back: Ethics and Politics after Deconstruction* (Albany: State University of New York Press, 2005).

12. "Philosophy in a Time of Terror," with Giovanni Borradori, in *Philosophy in a Time of Terror: Dialogues with Jurgen Habermas and Jacques Derrida* (Chicago: University of Chicago Press, 2003). He points out that the hijackers used US planes, US flight training facilities, and belonged to bin Laden's organization originally trained and financed by the United States. Moreover, these attacks were a response to perceived US economic and military activity.

13. Such as "Structure, Sign and Play in the Human Sciences," in *Writing and Difference*, trans. Alan Bass (London: Routledge, 1967).

14. Martin Heidegger, *Being and Time*, trans. Joan Stambaugh (Albany: State University of New York Press, 1996).

15. See "Critique of Pure Externalization," with Dr. Barbara Muraca, IAEP Annual Conference, Philadelphia, 2006.

16. This has led some to contest the use of the word "environment," as if it were just what surrounds us.

17. We invaded Iraq for its oil (2003), and we are entering the age of oil shortage. Yet an accelerating need for energy suggests that we will need to find cheap substitutes for oil, which may make the whole quest for oil dominance irrelevant.

18. World Resources Institute, World Conservation Union, *Global Biodiversity Strategy*, United Nations Environmental Programme, 1992.

19. For an excellent analysis of the conceptual and political incoherence of the ecofascism "smear," see David Orton, "Ecofascism: What Is It? A Left Biocentric Analysis," Green Web Bulletin #68, home.ca.inter.net/~greenweb/Ecofascism.html.

20. Operation Iraqi Freedom makes this clear.

21. See Heidegger's essay "Language" in *Poetry, Language Thought* (New York: Harper and Row, 1971), 210.

22. Jacques Derrida, *Of Grammatology* (Baltimore: Johns Hopkins University Press, 1976), 158.

23. The worry about deploying the value of freedom in this way obviously extends to military conquest and subordination of other peoples as well as to slavery and patriarchy, each of which have been justified in terms of subordinating a slave mentality to one that can exercise leadership.

24. I can imagine someone arguing that now that we have conquered genetic sequencing, we do not need actual other species anymore. If we ever need them, we can recreate them. Two objections: (1) if species disappear before we discover them, as mostly happens, then they will take their code with them into oblivion; the technology of gene mapping is *not* the same as life creation, so how would we know what life-forms to recreate? And (2) this presupposes sufficient continuing health of the planet to sustain these labs.

While it is not the focus of this section, I do not mean to neglect the importance of the life of each creature, quite apart from its species membership. Derrida once wrote that each death is the end of the world.

25. Is not human freedom an *essential* difference? Who is not tempted to say this? And yet we do need to ask: essential for what? For legitimating colonial, patriarchal, speciesist violence when we claim to be delivering it to our fellow men? Can we really separate our ontological intuitions from their political efficacy? And do we really know what we mean by freedom?

26. Jacques Derrida, *Writing and Difference*, trans. Alan Bass (Chicago: University of Chicago Press, 1978).

27. Steven M. Wise, in *Rattling the Cage: Toward Legal Rights for Animals* (New York: Perseus Books, 2000), argues the case for rights for animals as well as and legal personhood for chimps and bonobos.

28. Jacques Derrida, *The Other Heading*, trans. Pascale-Anne Brault and Michael B. Naas (Bloomington: Indiana University Press, 1992).

29. In this chapter, I mean by living beings both plants and animals, as well as all other creatures from single-celled organisms upward. I admit to mammalocentric tendencies, but I am fighting to overcome them. And I am assuming, too, that there are other proper objects of environmental concern, such as species survival, biodiversity, habitats, ecological health, and so forth.

30. We would also want to welcome the destabilizing of our assured sense of the human that we cannot avoid confronting in the writing of Nietzsche, for whom "man is a rope, tied between beast and overman—a rope over an abyss. . . . What is great in man is that he is a bridge and not an end." Friedrich Nietzsche, *Thus Spoke Zarathustra* (Harmondsworth: Penguin, 1961), Prologue 4. See also Donna Haraway's speculations in "A Cyborg Manifesto: Science, Technology, and Socialist-Feminism in the Late Twentieth Century," in *Simians, Cyborgs and Women: The Reinvention of Nature* (New York: Routledge, 1991), 149–181.

31. Jacques Derrida, "Force of Law: The 'Mystical Foundation of Authority,'" in *Deconstruction and the Possibility of Justice*, ed. Drucilla Cornell et al. (London: Routledge, 1992), 27.

32. The idea of a democracy-to-come is advanced, for example, in Derrida's *The Politics of Friendship* (London: Verso, 1997).

33. "Should Trees Have Standing?" (1972), in *Should Trees Have Standing? And Other Essays* (New York: Oceana, 1996).

34. Currently, for example, there seems to be a genuine conflict between two bodies representing the interests of trees (and offering certification of environmentally sustainable practices): the Sustainable Forestry Initiative and the Forest Stewardship Council. In fact, it soon becomes clear that the former actually represents the interests of paper and logging companies, while the latter attempts to represent the interests of the whole ecological space in which trees are grown.

35. Reference here to a parliament of living is a response to Bruno Latour's invocation of a parliament of things. I share some of his suspicion about the distinction between nature and culture, but I do think that living beings are the source of all *interest* and hence of what needs to be represented in a broader parliament. See Bruno Latour, *Politics of Nature: How to Bring the Sciences into Democracy*, trans. Catherine Porter

(Cambridge, MA: Harvard University Press, 2004). See also John Seed and Joanna Macy, *Thinking Like a Mountain: Towards a Council of All Beings* (Gabriola Island, BC, Canada: New Catalyst, 2007).

36. Derrida, *Writing and Difference*, 117.

37. See my "Giving Voice to Other Beings," in *Thinking Plant Animal Human* (Minneapolis: Minnesota University Press, 2019), Chapter 10. See also the interview with Tom Regan, author of *The Case for Animal Rights* (1983) (Berkeley: University of California Press, 2004): "As animal advocates, we have a reason to get up in the morning. A reason to rest at night. And that is to be a voice for the voiceless." See *Satya Magazine*, August 2004.

2. The Idea of Ecophenomenology

1. I argue elsewhere for the continuing need to explore the breakdown of the distinction between intentionality and causality. That treatment offered ways of talking about the "plexity" of time while dissolving "thinghood" into broader questions of identity and boundary maintenance. See "What Is Ecophenomenology?" in *The Step Back* (Albany: State University of New York Press, 2005), 149–68.

2. See, for example, Ted Toadvine in *Merleau-Ponty's Philosophy of Nature* (Evanston, IL: Northwestern University Press, 2009).

3. Bill McKibben, *The End of Nature* (New York: Random House, 2006).

4. Wood, "What Is Ecophenomenology?"

5. See Heidegger's foreword to Husserl's *The Phenomenology of Internal Time Consciousness* (1910), trans. J. Churchill (The Hague: Martinus Nijhoff, 1964).

6. Recall Samuel Beckett, *Imagine Dead Imagine* (London: Calder, 1966).

7. Relationality gets a lot of airtime in this volume, and for good reason. Identity is in all sorts of ways relationally constituted. Grasping this marks ontological progress, and yet it opens up a whole set of new questions and difficulties. Some forms of relationality are oppressive and toxic, and threaten our integrity, while others make it possible.

8. Think of William James's reference to the baby's experience of "one great blooming, buzzing confusion," in *Principles of Psychology* (New York: Dover, 2012), Chapter 13.

9. Tim Ingold, *Being Alive: Essays on Movement, Knowledge and Description* (London: Routledge, 2011), 6.

10. Ingold, *Being Alive*, 10.

11. Ingold, *Being Alive*, 10.

12. J. J. Gibson, *The Ecological Approach to Visual Perception* (1979) (New York: Routledge, 2014).

13. Ingold, *Being Alive*, 12.

14. Maurice Merleau-Ponty, "Eye and Mind," in *The Primacy of Perception*, ed. James E. Edie, trans. Carleton Dallery (Evanston, IL: Northwestern University Press, 1964).

15. Merleau-Ponty, "Eye and Mind," in Edie, *The Primacy of Perception*, 129.

16. Merleau-Ponty, "Eye and Mind," in Edie, *The Primacy of* Perception, 129.

17. Merleau-Ponty, "Eye and Mind," in Edie, *The Primacy of Perception*, 130.

18. Jane Bennett, *Vibrant Matter: A Political Ecology of Things* (Durham, NC: Duke University Press, 2010).

19. Mick Smith, *Against Ecological Sovereignty* (Minneapolis: University of Minnesota Press, 2011).

20. In Martin Heidegger, *Poetry Language Thought* (New York: Harper, 2013).

21. See Heidegger's "Letter on Humanism," in *Martin Heidegger: Basic Writings* (New York: Harper, 2008).

22. See Philippe Lacoue-Labarthe, "Transcendence Ends in Politics," in *Typography: Mimesis, Philosophy, Politics*, ed. Christopher Fynsk (Cambridge, MA: Harvard University Press, 1989).

23. Originally anonymous but written, it turns out, by Colonel Mark Mykelby and Captain Wayne Porter (2011).

24. We might, perhaps, try to understand the appeal of "science" in terms of the practical and constructive advantages that certainty, guaranteed foundations, would offer if they were actually available.

3. Ecological Imagination: A Whiteheadian Exercise in Temporal Phronesis

1. And with accelerationism, the idea that it might be speeding up. See, for example, Robin Mackay and Armen Avanessian, eds., *Accelerate: The Accelerationist Reader* (Falmouth: Urbanomic, 2014).

2. A. N. Whitehead, *Science and the Modern World* (Harmondsworth: Penguin, 1960), 105.

3. Whitehead, *Science and the Modern World*, 105.

4. See Martin Heidegger, "Letter on Humanism," in *Martin Heidegger: Basic Writings* (New York: Harper, 2009).

5. Whitehead, *Science and the Modern World*, 101.

6. Whitehead, *Science and the Modern World*, 103.

7. It is arguably an advance in the cause of "enlightenment" to realize that the very idea of freedom, for example, is problematic.

8. A. N. Whitehead, *Adventures of Ideas* (New York: Free Press, 1967), 255–256.

9. Whitehead, *Adventures of Ideas*, 246.

10. Whitehead, *Adventures of Ideas*, Chapter 4.

11. A. N. Whitehead, *Process and Reality* (New York: Free Press, 1979), 41.

12. Whitehead, *Adventures of Ideas*, 60.

13. A. N. Whitehead, *The Function of Reason* (Princeton, NJ: Princeton University Press, 1929).

14. See Friedrich Nietzsche: "To breed an animal with the right to make promises—is not this the paradoxical task that nature has set itself in the case of man?" (Essay 2 in *The Genealogy of Morals* [Oxford: Oxford University Press, 2009]).

15. See, for example, Jacques Derrida, *Limited Inc.*, ed. Gerald Graff, trans. Jeffrey Mehlman and Samuel Weber (Evanston, IL: Northwestern University Press), 1998.

16. See my *The Deconstruction of Time* (Evanston, IL: Northwestern University Press, 2001).

17. Whitehead, *Adventures of Ideas*, 243.

18. "The passage of nature which is only another name for the creative force of existence has no narrow ledge of definite instantaneous present within which to operate. Its operative presence which is now urging nature forward must be sought for throughout the whole, in the remotest past as well as in the narrowest breadth of any present duration. Perhaps also in the unrealized future. Perhaps also in the future which might be as well as the actual future which will be. It is impossible to meditate on time and the mystery of the creative passage of nature without an overwhelming emotion at the limitations of human intelligence" (A. N. Whitehead, *The Concept of Nature* [New York: Cosimo Classics, 2007], 73).

19. Whitehead, *The Function of Reason*, 19.

20. The rabbit was introduced from Europe to New Zealand in the mid-nineteenth century as game for sportsmen. Numbers quickly rose to plague proportions. Ferrets, stoats, weasels, and cats were introduced in an attempt to control the rabbits, with disastrous consequences for native bird life.

21. Whitehead, *The Function of Reason*, Chapter 3.

22. Whitehead, *Adventures of Ideas*, 60.

23. Whitehead, *The Function of Reason*, 12.

24. Whitehead, *Science and the Modern World*, 105.

25. Whitehead, *Science and the Modern World*, 103.

26. Whitehead, *The Function of Reason*, 8.

27. Whitehead, *The Function of Reason*, Chapter 3.

4. The Eleventh Plague: Thinking Ecologically after Derrida

1. See Martin Heidegger, *What is Called Thinking?* (1954) (New York: Harper, 1968), 77.

2. Jacques Derrida, *Specters of Marx: The State of the Debt, the Work of Mourning, and the New International* (New York: Routledge, 1994), 81–84.

3. I address these issues in a somewhat different way in "Derrida Vert?," *Oxford Literary Review* 36, no. 2 (2014): 319–22; "Specters of Derrida: On the Way to Econstruction," in *Ecospirit: Religions and Philosophies for the Earth*, ed. Laurel Kearns and Catherine Keller (New York: Fordham University Press, 2007), 264–290; and "Globalization and Freedom," in *The Step Back: Ethics and Politics after Deconstruction* (Albany: State University of New York Press, 2005).

4. John Keats, the poet. From a letter to one of his brothers (1817).

5. The classic paper is Lynn White, "The Historical Roots of Our Ecological Crisis," *JASA* 21 (June 1969): 42–47.

6. This chapter is a version of my paper in Matthias Fritsch, Philippe Lynes, and David Wood, eds., *Eco-Deconstruction: Derrida and Environmental Ethics* (New York: Fordham University Press, 2018). There is no essential difference between econstruction (Chapter 3) and what we have come to call ecodeconstruction.

7. See his Gifford Lectures, 2013, https://www.giffordlectures.org/lectures/facing-gaia-new-enquiry-natural-religion.

8. See William Cronon, ed., *Uncommon Ground: Rethinking the Human Place in Nature* (New York: W. W. Norton, 1995).

9. Bill McKibben, *The End of Nature* (New York: Random House, 2006).

10. I retain the word "man" here because of, not despite, its ideologically regressive legacy. As a reminder!

11. Naive versions of the Enlightenment project persist even today. See Steven Pinker, *Enlightenment Now: The Case for Reason, Science, Humanism, and Progress* (New York: Viking, 2018).

12. Jacques Derrida and Jurgen Habermas, *Le "Concept" du 11 septembre* (Paris: Galilée, 2004), translated as *Philosophy in a Time of Terror*, ed. Giovanna Borradori (Chicago: University of Chicago Press, 2004).

13. Derrida and Habermas, *Le "Concept" du 11 septembre*, translated as *Philosophy in a Time of Terror*.

14. Friedrich Nietzsche, "Truth and Falsity in their Ultra Moral Sense," in *The Complete Works of Friedrich Nietzsche* (London: T. N. Foulis, 1911).

15. The TV program *America's Most Wanted* uses names (and images) to apprehend criminals.

16. I pursue these ideas more systematically in *The Deconstruction of Time* (Evanston, IL: Northwestern University Press, 2001); *Time after Time* (Bloomington: Indiana University Press, 2007); and *Deep Time, Dark Times: On Being Geologically Human* (New York: Fordham University Press, 2018).

17. I have in mind here the shift from *Being and Time* (1926) to "Time and Being" (1962).

18. Jacques Derrida, *Of Grammatology* (Baltimore: Johns Hopkins University Press, 1976), 5.

19. See, for example, Sam Scheffler, *Death and the Afterlife* (Oxford: Oxford University Press, 2016).

20. One of the more remarkable comments made by a witness to the Skopes ("Monkey") trial (Dayton, Tennessee, 1925) was that he couldn't see what the fuss was about. He didn't mind being descended from monkeys. But fish? No way!

21. For more on this, see my *Deep Time, Dark Times*, and my *Thinking Plant Animal Human* (Minneapolis: University of Minnesota Press, 2019), esp. Chapter 1, "Homo Sapiens: The Long View."

22. Such as *Homo heidelbergensis*, *Homo rhodesiensis* or *Homo antecessor*, *Homo erectus*, *Homo denisova*, *Homo floresiensis*, and *Homo neanderthalensis*.

23. It is a tragic but sobering thought that this capacity for (self-) destruction might contribute to saving the planet. Genocide to the rescue? Gaia moves in mysterious ways? This is surely as obscene a thought as it is sobering.

24. Jacques Derrida, *Spectres de Marx* (Paris: Galilée, 2006), translated by Peggy Kamuf as *Specters of Marx* (New York: Routledge, 2006).

25. The anthropological machine is an expression coined by Giorgio Agamben in *The Open: Man and Animal* (Stanford: Stanford University Press, 2002), Chapter 9.

26. "Comment ne pas manger: Deconstruction and Humanism," Death of the Animal conference, Warwick, November 1993. Also Chapter 9 of *Thinking after Heidegger* (Cambridge: Polity, 2002).

27. Or, as Derrida says, "betise"! See *La Bête et le souverain, Volume I (2001-2002)* (Paris: Galilée, 2008), translated by Geoffrey Bennington as *The Beast and the Sovereign, Volume I* (Chicago: University of Chicago Press, 2009).

28. "Eating Well or the Calculation of the Subject," in *Who Comes after the Subject?*, ed. Eduardo Cadava, Peter Connor, and Jean-Luc Nancy (New York: Routledge, 2001). See also *Points de suspension: Entretiens* (Paris: Galilée, 1992), translated by Peggy Kamuf as *Points . . . Interviews, 1974-1994* (Stanford: Stanford University Press, 1995).

29. I introduce this term in "Kinnibalism, Cannibalism: Stepping Up to the Plate." https://www.academia.edu/6813639/Kinnibalism_Cannibalism_Stepping_Up_to_the.

30. While this can happen, I have argued forcibly against this line of thought. See "Comment ne pas manger: Deconstruction and Humanism," in *Thinking after Heidegger* (Cambridge: Polity, 2002), Chapter 9. Vegetarianism can just as easily be the leading edge of a broader transformation.

31. See Jacques Derrida, *The Animal That Therefore I Am* (New York: Fordham University Press, 2008).

32. See Elizabeth Kolbert, *The Sixth Extinction* (New York: Henry Holt, 2014). The title has a strange ambivalence to it. It announces the geological scale of what we are bringing about. And yet it is hardly unprecedented—there were five previous ones; the earth yawns.

33. See my discussion of agency in *Deep Time, Dark Times*, especially Chapters 6 and 7.

34. Ludwig Wittgenstein, *Philosophical Investigations* (London: Pearson, 1973), #43.

35. Martin Heidegger, *On Time and Being* (Chicago: University of Chicago Press, 2002), 2.

5. Things at the Edge of the World

1. See Martin Heidegger, *What Is a Thing?* (Chicago: Henry Regnery, 1968), and "The Origin of the Work of Art," in *Off the Beaten Track* (Cambridge: Cambridge University Press, 2002). See Hans Georg Gadamer, *Truth and Method*, trans. J. Weinsheimer and D. G. Marshall (New York: Crossroad, 1975).

2. The word "thing" is becoming something to be fought over. With an explicit negative reference to Heidegger, Bruno Latour appropriates it for political ends: "Long before designating an object thrown out of the political sphere and standing there objectively and independently, the Ding or Thing has for many centuries meant the issue that brings people together because it divides them." Bruno Latour, introduction to *Making Things Public: Atmospheres of Democracy*, ed. Bruno Latour and Peter Weibel (Cambridge, MA: MIT Press, 2005).

3. The extended book project *Things at the Edge of the World*, for which this essay serves as a trailer, outlines a truly fractal ontology.

4. An initial list of "things," which can be expanded very easily, included mouth, body, animal, tree, sun, painting, 9/11, God, death, woman, book, and earth.

5. What is implied by the thought that the life made possible by the constancy of the sun's energy can now think this?

6. It should be clear that I have little in common with Bataille's "general economics," for which the sun's endless supply of energy provides an excess we need to spend. Eroticism does not need such a cosmological economics. And global warming has changed the name of the wider game.

7. In about five billion years, when the sun will "eat" the earth as it expands and dies.

8. See Aldo Leopold, *Sand County Almanac* (London: Oxford University Press, 1968).

9. "To whom does this terrace belong? / With its limestone crumbling into fine greyish dust, / Its bevy of bees, and its wind-beaten rickety sunchairs. / Not to me, but this lizard, / Older than I, or the cockroach." From "The Lizard," *Collected Poems of Theodore Roethke* (London: Faber and Faber, 1968).

10. "We reached the dying in her eyes. . . . There was something new to me in those eyes—something known only to her and the mountains. I was young then, and full of trigger-itch. I thought that because fewer wolves meant more deer that no wolves would mean hunter's paradise, but after seeing the green fire die, I sensed that neither the wolf nor the mountain agreed with such a view." Leopold, *Sand County Almanac.*

11. Jakob von Uexküll, "A Stroll through the Worlds of Animals and Men," in *Instinctive Behavior: The Development of a Modern Concept,* ed. Claire Schiller (New York: International Universities Press, 1957).

12. Anthropology has a growing literature on this topic. A contemporary classic is Eduardo Kohn, "How Dogs Dream: Amazonian Natures and the Politics of Transspecies Engagement," *American Ethnologist* 34, no. 1 (2007): 3–24—"an appreciation for Amazonian preoccupations with inhabiting the points of view of nonhuman selves, to move anthropology beyond 'the human.'" (4). Through such studies, Westerners can expand their own capacity to imagine nonhuman perspectives.

13. Martin Heidegger, "What Is Metaphysics?," in *Martin Heidegger: Basic Writings,* ed. David Ferrel Krell (New York: HarperCollins, 1993).

14. I will explain the infinite as the absence of measure in a phenomenon in which an absolute distinction and an absence of concrete measure are laminated together.

15. See her essay "Sexual Difference," in *The Irigaray Reader,* ed. Margaret Whitford (Cambridge: Basil Blackwell, 1991).

16. I explore this line of criticism in "Where Levinas Went Wrong: Some Questions for My Levinasian Friends," in *The Step Back* (Albany: State University of New York Press, 2005).

17. See Gadamer, *Truth and Method,* and "The Relevance of the Beautiful," in *The Relevance of the Beautiful and Other Essays,* trans. Nicholas Walker (New York: Cambridge University Press, 1986).

18. I explore these themes of existential temporal constitution with respect to Heidegger in "Reading Heidegger Responsibly: Glimpses of Being in Dasein's Development," in *Heidegger and Practical Philosophy,* ed. François Raffoul and David Pettigrew (Albany: State University of New York Press, 2002).

19. The Production Code, also known as the Hays Code, was the set of industry censorship guidelines that governed the production of US films from 1930 to 1968.

20. See Kelly Oliver, *Knock Me Up, Knock Me Down: Images of Pregnancy in Hollywood Films* (New York: Columbia University Press, 2012).

21. Merleau-Ponty and I agree on the importance of a recurrent feature that is both structure and event, one whose boundaries are still to be negotiated. Sometimes, for example, it seems we are talking about a short list of privileged things that have an exemplary reversibility operator status. At other times, it seems that anything, suitably appreciated, can take on this role. (Compare Heidegger talking about great works of art.) What, then, is a thing? In another departure from Merleau-Ponty, my list of things radically exceeds the sensible. The inclusion of death makes this clear, where the double movement—both opening up (if you like) a world of meaning and facilitating a certain deconstruction of selfhood—is clearly in play.

22. The phrase comes from Winnicott, who studied with Klein. "The first ego organization comes from the experience of threats of annihilation which do not lead to annihilation and from which, repeatedly, there is recovery." Donald Winnicott. "Primary Maternal Preoccupation," in *Collected Papers: Through Paediatrics to Psycho-Analysis* (London: Tavistock, 1956).

23. Plato, *Theaetetus and Sophist* (Cambridge: Cambridge University Press, 2012), *Sophist* 244a.

24. Needless to say, agency is not dependent on (and may indeed preclude) an exaggerated sense of autonomy, and a "deconstruction" of the myth of the autonomous subject does not seek to abolish the agent-subject but to make its constitutive relationality visible and productive. See, for instance, Judith Butler, *Bodies That Matter* (New York: Routledge, 1993).

25. See Nelson Goodman, *Ways of Worldmaking* (New York: Hackett, 1978).

26. The revelations of torture at Abu Ghraib prison left many people asking, "Is that our America?"

27. Is the deployment of "strangeness" here not itself highly anthropocentric? I think it is, and I am indeed defending a certain distinctiveness of the human. Any creature that felt cosmic wonder could join the club. This should not close off wondering what (analogues of) wonder/horror/disgust might look like for other animal kinds. Chimps, dogs, and elephants—just to begin the list—act very "strangely" when confronted with death.

28. Martin Heidegger, *Introduction to Metaphysics* (New Haven, CT: Yale University Press, 2000), 123.

29. Heidegger, *Introduction to Metaphysics*, 133.

30. This is an allusion to Simon Critchley's *Infinitely Demanding* (London: Verso, 2007).

6. Reversals and Transformations

1. See "The Return of Experience," in my *Thinking after Heidegger* (Oxford: Polity, 2002), Chapter 2.

2. Reversal, repetition, and return are neither synonyms nor strictly dissociable. Moreover, there is a wider world of the re- that would embrace more loosely connected family members: reworking, remarks, reflection, reanimate, revelation, and perhaps even religion. What is key is the temporally articulating play of sameness and difference, in which it becomes clear that the force of inaugural events can only be retained by their transformed restaging, that identity is not so much presupposed by repetition as made possible by it, and that structural or schematic reversals both lay bare constitutive orderings of the real and enable its transformation. In the world of the re-, it is not the circle that is king—it is the spiral.

3. For Hume, experience was interrogated as the source for both impressions and ideas, and his reductive analysis of how we come up with the beliefs that set real limits to what we can seriously claim to know, corrosively undermining our grasp of both self and the "external" world. In his celebrated example of a reported miracle, he suggests it more likely that the witness is in error, given that miracles, by definition, violate established laws of nature. He does not seem to consider the case in which one witnesses the miracle oneself. One can come to doubt one's memory, but at the time?

4. Michel Foucault, "The Discourse on Language," in *The Archaeology of Knowledge* (New York: Pantheon, 1972), 236.

5. Martin Heidegger, "Origin of the Work of Art," in *Poetry, Language, Thought* (New York: Harper, 1970), 79.

6. In other words, to speak of an experience is all too easily to attribute to it, and to the subject "having" the experience, a primitive unity. Learning from experience, on that model, would be a deepening process, not one for which experience would be a disruption.

7. Jacques Derrida, "Force of Law: The Mystical Foundation of Authority," in *Deconstruction and the Possibility of Justice*, ed. Drucilla

Cornell, Michel Rosenfeld, and David Gray Carlson (New York: Routledge 1992), 24.

8. Jacques Derrida, *Aporias* (Stanford, CA: Stanford University Press, 1993), 93.

9. Ludwig Wittgenstein, *Tractatus Logico-Philosophicus* (New York: Dover, 1998), 6.431; emphasis original.

10. William James, *Varieties of Religious Experience* (Harmondsworth: Penguin, 1982). On a vacation climb in the Adirondacks in 1898, James himself underwent a variety of religious experiences. In a letter to his wife, he writes: "It seemed as if the Gods of all the nature-mythologies were holding an indescribable meeting in my breast with the moral Gods of the inner life. Doubtless in more ways than one, things in the Edinburgh lectures will be traceable to it" (preface).

11. Martin Heidegger, "Language," in *Poetry, Language, Thought* (New York: Harper, 1970), 210.

12. Jacques Derrida, *Of Grammatology* (Baltimore: Johns Hopkins University Press, 1976).

13. Ludwig Wittgenstein, "On Heidegger on Being and Dread," in *Heidegger and Modern Philosophy*, ed. Michael Murray (New Haven, CT: Yale University Press, 1978).

14. "To write poetry after Auschwitz is barbaric. And this corrodes even the knowledge of why it has become impossible to write poetry today." Theodor Adorno, "An Essay on Cultural Criticism and Society," in *Prisms* (Princeton, NJ: Princeton University Press, 1983), 34.

15. Monica Wittig, *The Lesbian Body* (New York: Beacon, 1986).

16. See Dylan Thomas, *The Collected Poems of Dylan Thomas, 1934-1952* (New York: New Directions, 1971).

17. See Friedrich Nietzsche, "Truth and Falsity in their Ultra-Moral Sense," in *The Existentialist Tradition: Selected Writings*, ed. Nino Langiulli (Garden City, NY: Anchor Books, 1971), and Blaise Pascal, *Pensées* (Harmondsworth: Penguin, 1995).

18. Rainer Maria Rilke, *Duino Elegies* (Portland, OR: Tavern Books, 2013), i.

19. Max Scheler, *The Nature of Sympathy* (London: Routledge, 2008).

20. Aldo Leopold, *A Sand County Almanac* (Oxford: Oxford University Press, 1968).

21. Leopold, *A Sand County Almanac*.

22. D. H. Lawrence, "The Snake," in *Selected Poems* (Harmondsworth, Penguin, 1989), 134.

23. Jacques Derrida, *The Animal That I Am* (New York: Fordham University Press, 2008).

24. See Arthur O. Lovejoy, *The Great Chain of Being* (Cambridge, MA: Harvard University Press, 1976).

25. This was one of Nietzsche's repeated moves: on women (neither deep nor shallow) and on truth (neither appearance nor reality).

26. See Derrida, *The Animal That I Am.*

27. It need not. It could have been created for us to marvel at rather than to exercise dominion over.

28. This issue is taken up masterfully in Andrew Mitchell's *The Fourfold* (Evanston, IL: Northwestern University Press, 2015).

29. A reference to Lacan's account of the child's alienating development of a self through identification with an image of itself in a mirror.

30. Jean-Paul Sartre, *Critique of Dialectical Reason* (London: Verso, 2010).

31. *Excessance* tries to capture the way in which things happen that overflow what one might think were their conditions of possibility. Or again, the way in which the real overflows our words, images, or other constructions. Cf. "There are more things in heaven and earth, Horatio, than are dreamt of in your philosophy" (William Shakespeare, *Hamlet*).

32. Edmund Husserl, *Formal and Transcendental Logic* (Dordrecht: Nijhoff, 1977).

33. Maurice Merleau-Ponty, *The Visible and the Invisible* (Evanston, IL: Northwestern University Press, 1968).

34. In Herbert Marcuse, *Eros and Civilization* (New York: Beacon, 1955).

35. But, just as Nietzsche described the possibility of horror at the revelation of the eternal return, one could react in a parallel way to what I am proposing. The world needs, as Nietzsche himself put it on another occasion, a yes, a no, a straight line. It needs creators who do not question their own genius, who have little time for introspection. They have a world to make or to conquer. The world is already overfull of the tender-minded, as James put it. It needs the tough minded, people willing to make hard decisions, even in the face of the suffering these decisions cause. Kierkegaard's account of despair, Sartre's description of nausea, or Heidegger's discussion

of angst or dread are self-indulgent forms of escapism, intellectual moaning minnies, serving only to weaken the capacity for individual agency already under threat by consumer culture and the corporatization of the world. Many could agree with these sentiments unreservedly when it comes to art. The risk a writer or painter or musician takes largely affects only a small circle. And the material they deploy—paint, words, sound—will never be in short supply. But worldly decisiveness on a grand scale can bring both salvation and devastation. A climate change dictator might indeed save the planet. But she could not do so without effecting, directly or indirectly, a more modest manner of dwelling. Decisiveness in fossil fuel and environmental deregulation, however, will spell disaster. The unbridled creativity we may affirm in art does not translate unrestrictedly into public policy.

7. Touched by Touching: Toward a Carnal Hermeneutics

1. Jacques Derrida, *On Touching—Jean-Luc Nancy* (Stanford, CA: Stanford University Press, 2005), 75.

2. Richard Kearney, "Diacritical Hermeneutics," in *Hermeneutic Rationality: La rationalité herméneutique*, ed. Maria Luísa Portocarrero, Luis António Umbelino, and Andrzej Wiercinski (Berlin: Lit Verlag Dr. W. Hopf, 2012), 177.

3. Needless to say, these reflections bear the trace of having brushed up against Derrida's magisterial *On Touching—Jean-Luc Nancy* (2005) and various texts written by Nancy to which he refers, especially *The Sense of the World* (1993) and *Corpus* (2008). If I do not engage substantially with these writings, I have indeed been moved by them.

4. The reference here is to Derrida's *The Animal That Therefore I Am* (New York: Fordham University Press, 2008).

5. My attention to chance sentences that land at one's feet as if on scraps of paper blown by the wind finds some resonance with Derrida's fastening onto an inscription on a Paris wall: "Quand nos yeux se touchent, fait-il jour ou fait-il nuit?"/ "When our eyes touch is it day or is it night?" (*On Touching*, 4), or, again: "Someone, you or me, comes forward and says: I would like to *learn* to *live* finally" (from the Exordium to *Specters of Marx* [1994], xvii; emphasis original). These words come to us. What are we to make of them?

6. George Berkeley, *An Essay towards a New Theory of Vision* (1703) §44 (Google digitized edition, 2008).

7. Friedrich Nietzsche, *Gay Science* (1882), trans. Walter Kaufmann (New York: Random House, 1991), §59.

8. Martin Heidegger, *Being and Time*, trans. Joan Stambaugh (Albany, State University of New York Press, 1996), 129.

9. From Richard Howard, "A Note on the Text," in Roland Barthes, *The Pleasure of the Text* (New York: Hill and Wang, 1975), vii.

10. Pablo Neruda, *Selected Poems in Translation*, online edition (A. S. Kline, 2000).

11. Gerard Manley Hopkins, "Pied Beauty," in *Gerard Manley Hopkins: Poems and Prose* (Harmondsworth: Penguin, 1985), 30.

> GLORY be to God for dappled things—
>> For skies of couple-colour as a brinded cow;
>>> For rose-moles all in stipple upon trout that swim;
>> Fresh-firecoal chestnut-falls; finches' wings;
>>> Landscape plotted and pieced—fold, fallow, and plough;
>>> And áll trádes, their gear and tackle and trim.
>> All things counter, original, spare, strange;
>>> Whatever is fickle, freckled (who knows how?)
>>> With swift, slow; sweet, sour; adazzle, dim;
>> He fathers-forth whose beauty is past change:
>>> Praise him.

12. In von Uexküll's words, "the whole rich world around the tick shrinks and changes into a scanty framework consisting, in essence, of three receptor cues and three effector cues—her Umwelt. But the very poverty of this world guarantees the unfailing certainty of her actions, and security is more important than wealth. By abridging her indulgence in sensual luxuries, nature has ensured the tick as failsafe an existence as the natural world provides. For animals like us, bathed continually in our own set of sensory inputs, only our imaginations allow us entry into the world of such a creature, an entry essential to the ethological perspective." Jakob von Uexküll, *A Foray into the Worlds of Animals and Humans: With a Theory of Meaning* (Minneapolis: University of Minnesota Press, 2010).

13. See Derrida, glossing Nancy: "All one ever does touch is a limit. To touch is to touch a limit, a surface, a border, an outline." *On Touching*, 103.

8. My Place in the Sun

1. "Having a place means that you know what a place means . . . what it means in a storied sense of myth, character, and presence but also in an ecological sense . . . integrating native consciousness with mythic consciousness" (Gary Snyder).

2. Ed Casey, *The Fate of Place: A Philosophical History* (Berkeley: University of California Press, 1998); Ed Casey, *Getting Back into Place* (Bloomington: Indiana University Press, 2009).

3. Even German colonialism made use of this sense, however disingenuously: Foreign Minister Bernhard von Bülow used it to launch Germany's policy of *Realpolitike*. In a parliamentary debate, he argued that "[i]n one word: we do not want to place anyone into the shadow, but we also claim our place in the sun." Parliamentary debate, December 6, 1897.

4. Emmanuel Levinas, "Ethics as First Philosophy," in *The Levinas Reader*, ed. Sean Hand (Oxford: Blackwell, 1997), 82.

5. John Locke, *Second Treatise on Government* (Cambridge: Cambridge University Press, 1988), sect. 27.

6. Consider Ezra Pound's question: Would you prefer an empty cell or one with an open sewer running through it? And stories of prisoners befriending the cell mouse. Anything to establish relation.

7. If a home is essentially a shared space, and a basic right, solitary confinement would be a violation of such a right, facilitated perhaps by a confusion between space and place.

8. Things get more problematic with Borges's discussion of exactly rewriting Cervantes's *Don Quixote* as a quite different book. These issues become more concrete when one thinks of intellectual property rights to popular songs, for example.

9. This criteria could be and was used to dispossess indigenous people of their land in Australia and in North America when their way of owning territory—nomadic, for example—did not meet the "mixing labor" specification.

10. It is rare but not unheard of for parties stranded in remote locations to eat their dead companions. Whether or not this is legal, it is understandable. Killing those who would otherwise be dead tomorrow for dinner today is equally understandable in the circumstances, even though it is hard to imagine a legal system that would condone it.

11. It goes without saying that these questions go right to the heart of the Palestinian "right to return."

12. See, for example, George Santayana, *The Realms of Being*, vol. II, *The Realm of Matter* (New York: Charles Scribner's Sons, 1942). I owe this reference to my colleague John Lachs.

13. It is worth noting how irresistible spatial analogies are. Bergson, noted for lamenting our tendency to think time spatially, writes of memory as an inverted cone, a pyramid.

14. A common book title. See, for instance, Thomas Huxley (1863), *Man's Place in Nature* (New York: Modern Library, 2001), and Pierre Teilhard de Chardin (1966), *Man's Place in Nature* (New York: Harper Collins, 2000). Each advocates an evolutionary approach, with the latter extending evolution into a spiritual dimension, and speculates on the future. Charles F. Hockett, in his *Man's Place in Nature* (Columbus: McGraw-Hill, 1973), offered an account of the overlap between human and animal linguistic capacity. While many components are variously shared with nonhumans, only humans have the full complement of linguistic capacities.

15. Reuven Firestone, "Jerusalem: Jerusalem in Judaism, Christianity, and Islam," in *Encyclopedia of Religion*, 2nd edition (Farmington Hills, MI: Thomson Gale, 2005), 4838. This article is a brilliant account of the various sanctifications of Jerusalem in its service to different religions.

16. These quotes are not cited for their historical accuracy (or otherwise) but for their exemplary status in explaining how Jerusalem can be claimed by all sides.

17. http://www.thefreedictionary.com/Jerusalem.

18. http://www.jewishvirtuallibrary.org/jsource/vie/Jerusalem1.html (page no longer active).

19. http://www.goisrael.com/Tourism_Eng/Tourist+Information/ Discover+Israel/Cities/Jerusalem.htm (page no longer active).

20. http://en.wikipedia.org/wiki/Religious_significance_of_Jerusalem.

21. See Part I of his *Groundwork of the Metaphysics of Morals* (Cambridge: Cambridge University Press, 2012).

22. Jakob Huber, "Cosmopolitanism for Earth Dwellers: Kant on the Right to be Somewhere," *Kantian Review* 22, no. 1 (2017): 1–25.

9. On Being Haunted by the Future

1. Jacques Derrida, *Specters of Marx* (London: Routledge, 1994), 64–65.

2. Michael McCarthy, "Global Warming: Passing the 'Tipping Point,'" *The Independent*, February 11, 2006.

3. See his *Margins of Philosophy*, trans. Alan Bass (Chicago: University of Chicago Press, 1984).

4. This move is particularly apparent in his *Contributions* and is not unrelated to his disillusionment with the Nazis. See *Contributions to Philosophy (From Enowning)*, trans. Parvis Emad and Kenneth Maly (Bloomington: Indiana University Press, 1999).

5. Michael Sprinkler, ed., *Ghostly Demarcations: A Symposium on Jacques Derrida's* Specters of Marx (London: Verso, 1999), 249.

6. Jacques Derrida and Elisabeth Roudinesco, *For What Tomorrow: A Dialogue* (Stanford, CA: Stanford University Press, 2004), 53.

7. Sprinkler, ed., *Ghostly Demarcations*, 249.

8. Walter Benjamin, Illuminations (New York: Schocken, 1969), 263.

9. Jacques Derrida, *Ghostly Demarcations: A Symposium on Jacques Derrida's* Specters of Marx (London: Verso, 1999).

10. Derrida, *Ghostly Demarcations*.

11. Compare "speech is a kind of writing" and Derrida's claim that those who would denounce apocalyptic discourse (e.g., Kant) are not immune to it.

12. Quoted by Roudinesco in *Ghostly Demarcations*, ix.

13. "Good Morning America," September 1, 2005.

14. Condoleezza Rice, at that time national security advisor, was talking about 9/11 shortly after the event. In fact, when the 9/11 Commission probed into why the country was so ill-prepared for a terrorist attack, they discovered that the Federal Aviation Administration had been investigating for years the possibility that terrorists might hijack a plane and use it as a missile and had specifically warned airports long before 9/11.

15. Dan Robinson, *Voice of America* (radio), October 2004.

16. Congresswoman Nita Lowey, senior Democrat on the House Foreign Operations Subcommittee, reported by Robinson, *Voice of America*.

17. See, for example, his *One World: The Ethics of Globalization* (New Haven, CT: Yale University Press, 2002). It is noteworthy that, like Derrida, he calls the United States a "rogue nation."

18. Jacques Derrida, *The Politics of Friendship* (London: Verso, 2005).

19. See, for example, Giovanna Borradori, *Philosophy in a Time of Terror: Dialogues with Jurgen Habermas and Jacques Derrida* (Chicago: University of Chicago Press, 2003).

20. Here we need to consider questions of accountability in light of common assumptions about "discount rates," which make redundant all sorts of planning beyond seven to ten years in the future.

21. ExxonMobil funds some forty think tanks and other individuals to propagate global warming skepticism and misinformation. Republican strategist Frank Luntz is candid: "Doubt is our currency."

22. Phil Lesley, "Coping with Opposition Groups," *Public Relations Review* 18 (1992): 331.

23. Al Gore, in a public lecture at Vanderbilt University on global warming (spring 2006), claimed that a survey of eight-hundred-plus articles on global warming in science journals turned up none that expressed doubt about the phenomenon, while in a similar number of articles in popular magazines, more than 50 percent recorded debate among scientists.

24. See, for example, George Lakoff, *Don't Think of an Elephant! Know Your Values and Frame the Debate* (White River Junction, VT: Chelsea Green Publishing, 2004).

25. Ron Suskind, "Without a Doubt," *New York Times*, October 17, 2004, late edition-final, sec. 6, p. 44, col. 1.

26. Jacques Derrida, "History of the Lie," in *Futures: Of Jacques Derrida*, ed. Richard Rand (Stanford, CA: Stanford University Press, 2001), 94.

27. Derrida, "History of the Lie," in Rand, *Futures*, 95.

28. Derrida, "History of the Lie," in Rand, *Futures*, 93.

29. Lewis Lapham, "The Case for Impeachment: Why We Can No Longer Afford George W. Bush," *Harper's Magazine*, March 2006, 28, 29; Derrida, "History of the Lie," in Rand, *Futures*, 94.

30. Derrida, "History of the Lie," in Rand, *Futures*, 94.

31. Derrida, "History of the Lie," in Rand, *Futures*, 96–97.

32. Derrida, "History of the Lie," in Rand, *Futures*, 97.

33. Derrida, "History of the Lie," in Rand, *Futures*, 90.

34. *Rebuilding America's Defenses: Strategy, Forces and Resources for a New Century*, PNAC Report of September 2000, http://www.visibility 911.org/wp-content/uploads/2008/02/rebuildingamericasdefenses.pdf.

35. The report is dated February 2004. Natural Resources Defense Council website, January 6, 2006.

36. This undoubtedly makes silent reference to the ten plagues visited on the Egyptians in the book of Exodus to encourage them to release the Israelites from slavery: (1) blood, (2) frogs, (3) gnats, (4) flies, (5) disease in livestock, (6) boils, (7) hail, (8) locusts, (9) darkness, and (10) death of firstborn.

37. "This International has today the figure of suffering and of compassion for the ten plagues of global order I enumerate in *Specters of Marx*. It decries that of which one speaks so little in the official political

rhetoric and in the discourse of 'engaged intellectuals,' even among the declared champions of human rights. To give some examples of easily distracting macro-statistics, I think of the millions of children who drown every year, of the nearly 50 per cent of women who are beaten or fall victim to sometimes murderous abuse (the 60 million disappeared women, the 30 million mutilated women), of the 23 million infected with AIDS (of which 90 per cent are in Africa and to whom the budget of AIDS research dedicates only 5 per cent of its resources, while therapy remains unavailable outside small occidental milieus), I think of the selective infanticide of girls in India and of the monstrous conditions of child labor in many countries, and of the fact that there are, I believe, a billion illiterate people and 140 million uneducated children, I think of the maintenance of the death penalty and of the circumstances of its administration in the United States (the only Western democracy to do so and a country that no longer recognizes the convention concerning children's rights and continues to execute punishment against minors even after they have reached adult age, etc.). I quote these numbers, published in official reports, from memory in order to convey an idea of the scale of the problems that call for an 'international' solidarity of which no state, no party, no syndicate, no civic organization really takes charge. All who suffer and all those who are not insensitive to the dimension of these urgent issues belong to this International, everybody who—civic or national background notwithstanding—is determined to draw the attention of politics, law and ethics towards them." Interview with Thomas Assheuer in *Die Zeit*, March 5, 1998.

38. Derrida made this comment orally in reply to my paper "Globalization and Freedom," presented at the "Returns of Marx" conference, Paris, March 2003.

39. See interview with Thomas Assheuer in *Die Zeit*, March 5, 1998.

40. For references to going through the undecidable as a condition of responsibility and justice, see, for example, "Force of Law": "A decision that didn't go through the ordeal of the undecidable would not be a free decision, it would only be the programmable application or unfolding of a calculable process. It might be legal; it would not be caring [juste]" (Jacques Derrida, "Force of Law: The Mystical Foundation of Authority," trans. Mary Quaintance, *Cardozo Law Review* 11 [1990]: 963). Jacques Derrida, *Rogues: Two Essays on Reason* (Stanford, CA: Stanford University Press, 2005), 97.

41. A distinction drawn first by Windelband to mark the distinction between historical and natural sciences ("Geschichte und Naturwissenschaft: Straßburger Rektoratsrede," in Wilhelm Windelband, *Präludien: Aufsätze und Reden zur Philosophie und ihrer Geschichte* [Tubingen: J. C. B. Mohr, 1894], 136–160).

42. Franz Kafka, quoted by Walter Benjamin, *Illuminations* (New York: Schocken, 1969), 116.

43. Martin Heidegger, *Introduction to Metaphysics*, trans. Gregory Fried and Richard Polt (New Haven, CT: Yale University Press, 2000).

44. Borradori, *Philosophy in a Time of Terror*, 98, 45.

45. Borradori, *Philosophy in a Time of Terror*, 116.

46. Borradori, *Philosophy in a Time of Terror*, 118.

47. Francis Fukuyama, *The End of History and the Last Man* (London: Macmillan, 1992).

48. Derrida, *Rogues*, 97.

49. *Rebuilding America's Defenses.*

50. *Rebuilding America's Defenses.*

51. Borradori, *Philosophy in a Time of Terror*, 93, 52.

52. Borradori, *Philosophy in a Time of Terror*, 113.

53. *Harrap's New Shorter French and English Dictionary* (London: Harrap, 1971).

54. Immanuel Kant, "On a Newly Arisen Superior Tone in Philosophy," in *Raising the Tone of Philosophy: Late Essays by Immanuel Kant, Transformative Critique by Jacques Derrida*, ed. Peter Fenves (Baltimore: Johns Hopkins University Press, 1993), 51–81. On Foucault, see "Cogito and the History of Madness," in *Writing and Difference*, trans. Alan Bass (Chicago: University of Chicago Press, 1978), 31–63.

55. See, for example, *Rebuilding America's Defenses.*

56. Derrida and Roudinesco, *For What Tomorrow*, 98, 57.

57. Borradori, *Philosophy in a Time of Terror*, 115.

58. Benjamin, *Illuminations*, 264.

59. Benjamin, *Illuminations*, 265.

10. Beyond Narcissistic Humanism: Or, in the Face of Anthropogenic Climate Change, Is There a Case for Voluntary Human Extinction?

1. What kind of resistance is reflected in Heidegger's reference to "our appalling bodily kinship with the animal"?

2. Jacques Derrida, *The Other Heading: Reflections on Today's Europe*, trans. Pascale-Anne Brault and Michael B. Naas (Bloomington: Indiana University Press, 1992).

3. "'Tis not contrary to reason to prefer the destruction of the whole world to the scratching of my finger. 'Tis not contrary to reason for me to chuse my total ruin, to prevent the least uneasiness of an Indian or person unknown to me." David Hume, *The Natural History of Religion*, ed. H. E. Root (Stanford, CA: Stanford University Press, 1967), 416.

4. Emmanuel Levinas, *Otherwise Than Being or Beyond Essence*, trans. Alphonso Lingis (The Hague: Martinus Nijhoff, 1981). More directly, and this is a pressing question with regard to Heidegger's treatment of "the animal," the question raised by Derrida, Llewelyn, and others is whether in effect Heidegger's entire strategy of the displacement of man toward his openness to being is not a covert vehicle for entrenching traditional humanistic preferences. Dasein, he will say, is world-forming, while the animal is "poor in world" and the rock altogether "without world." The critical response would be that despite his best efforts, Heidegger reproduces a quite traditional chain-of-being argument about the relative place of man, nonhuman life, and inert matter. And surely, in some respect this verdict is justified. But it would leave open the question as to whether Heidegger had not, through his ontological displacement, brought out something objectively important about a certain humanism. But what would allow us to conclude that Heidegger (or any other similar thinker) was elucidating the truth of a certain anthropocentrism rather than reproducing its unjustified privilege in a more sophisticated language?

5. See the Voluntary Human Extinction Movement, whose slogan is "May we live long and die out."

6. "Give me liberty, or give me death!" is a quotation attributed to Patrick Henry from a speech he made to the Second Virginia Convention on March 23, 1775, at St. John's Church in Richmond, Virginia. https://en.wikipedia.org/wiki/Give_me_liberty,_or_give_me_death!

7. "We must hang together, gentlemen . . . else, we shall most assuredly hang separately." Attributed to Benjamin Franklin.

8. See Martin Heidegger, *Country Path Conversations*, trans. Bret W. Davis, (Bloomington: Indiana University Press, 2010).

9. James Gustave Speth, *The Bridge at the Edge of the World: Capitalism, the Environment, and Crossing from Crisis to Sustainability* (New Haven, CT: Yale University Press, 2009). Speth was the dean of forestry

at Yale. He writes, "[M]ost environmental deterioration is a result of systemic failures of the capitalism that we have today and that long-term solutions must seek transformative change in the key features of this contemporary capitalism. The book addresses these basic features of the capitalism that we know, in each case seeking to identify the transformative changes needed. The good news is that impressive thinking and some exemplary action have occurred on the issues at hand. Proposals abound, many of them very promising, and new movements for change, often driven by young people, are emerging. These developments offer genuine hope and begin to outline a bridge to the future. The market can be transformed into an instrument for environmental restoration; humanity's ecological footprint can be reduced to what can be sustained environmentally; the incentives that govern corporate behavior can be rewritten; growth can be focused on things that truly need to grow and consumption on having enough, not always more; the rights of future generations and other species can be respected. But we haven't got much time." "Between Two Worlds," online interview, http://www.rachel.org/?q=es/node/6956.

10. Speth, *The Bridge at the End of the World*.

11. See also Naomi Klein, *This Changes Everything: Capitalism v. the Climate* (New York: Simon and Schuster, 2015).

12. Here let me say that I am drawing on another project, *Things at the Edge of the World*, in which I explore at length a broad range of experiences of reversal, through which we find ourselves not just subjects of experience but taken out of ourselves into a fractal world of multiply distributed sites in which we lose ourselves. In each case, a thing that begins as an object of experience becomes the site of an event of reversal and transformation in which not only am I implicated in an unexpected way, but the rest of the world is poised for restructuration, and the proliferation of new chains of possibility.

13. I explore this line of criticism in "Where Levinas Went Wrong: Some Questions for My Levinasian Friends," in *The Step Back* (Albany: State University of New York Press, 2005).

14. See, for example, Philippe Lacoue-Labarthe, "Transcendence Ends in Politics," in *Typography, Mimesis, Philosophy, Politics*, ed. Christopher Fynsk (Stanford, CA: Stanford University Press, 1989).

15. Luce Irigaray, *An Ethics of Sexual Difference*, trans. Carolyn Burke and Gillian C. Gill (Ithaca, NY: Cornell University Press, 1993), 171.

16. Informal correspondence.

17. Lyotard, for example, argues that the Enlightenment took us straight to the gas chambers.

18. This is very much Aldo Leopold's position: "To keep every cog and wheel is the first precaution of intelligent tinkering." See "The Round River," in *A Sand County Almanac: With Essays on Conservation from Round River* (New York: Ballantine Books, 1970). I am grateful to Laura Westra for drawing my attention to this.

19. Could we call this "transcendence"? I have argued that we cannot claim (and hence should not fool ourselves by believing we deserve) an exalted place on the planet on the basis of distinctive characteristics whose virtues we leave unfulfilled or undeveloped. This has direct political implications if we believe that humans are smart enough to know how to avoid catastrophic climate change but unable to muster the collective will to make that happen. Are we rational beings? There is indeed something extraordinary about what we call consciousness and freedom. But as Schelling demonstrated, with individual freedom comes the very real possibility of evil, the thrusting forth of the dark side, the assertion of the individual will in the face of the broader good, an analysis that arguably applies directly to our position as a species. Importantly, too, this suggests that "evil" does not just reflect some sort of lack (e.g., ignorance or blindness—it may be the corollary of something we genuinely and rightly value (creativity, shaping the natural world, replacing back-breaking human labor, etc.).

Index

9/11, 133, 243n14; Derrida on, 35, 88,
91, 100, 181, 189–90, 197, 224n12;
temporality of, 91

Abu Ghraib prison, torture at, 235n26
accelerationism, 228n1
Adorno, Theodor, 126–27, 237n14
Against Ecological Sovereignty (Smith),
62–64
Agamben, Giorgio, 232n25
agency: Bennett on, 61–62; constitutive
relationality and, 235n24; Derrida on,
41; Heidegger on, 20–22, 41
America's Most Wanted, 231n15
anamnesis, Plato and, 6
angst, 7; Heidegger on, 126
animal rights, 95; anthropomorphism
and, 87; democracy-to-come and,
46–49, 96, 226nn34,35; environmentalism versus, 32–33; naming and,
88–89; parliament of the living and,
46–49, 226n35; Wise on, 225n27
animals, 215, 230n20; carnal hermeneutics and, 143–45; Derrida on, 32–34,
89, 94–97, 117–18, 131, 143, 150–51,
212; empathy for, 129; experience
and, 240n12; Heidegger on, 22, 109–
10, 246n1, 247n4; Leopold and, 108,
130–31; reversal, experiences of, and,
105, 108–10, 130–32, 139–40, 212,
234nn10,12; Rilke on, 22, 109–10;
Roethke and, 108, 233n9; territories

and, 161; touch and, 143–44, 150–52;
Wittgenstein on, 144; wonder and,
235n27. *See also* nonhuman life-
forms
Anthropocene, 93–94
anthropocentrism: biocentric perspective on, 6, 203; Derrida on, 43–44;
enlightened, 6–7, 41–42, 49, 209,
216–17; Heidegger and, 208–10,
247n4; human cognitive capacities
and, 202–4, 214–16, 249n19; interdependency of organisms and, 203–4,
213–16; as logically unavoidable,
205, 208; mammalocentrism versus,
205, 207; narcissism and, 211, 216;
self-interest transcended by, 205,
208; solidarity, human, and, 206, 218;
strangeness and, 235n27; unabashed,
205–7; value, as exclusive to living
beings, and, 207; vulgar, 208–9
anthropological machine, 94, 232n25
anthropomorphism, animal rights and,
87
antibiotics, 97
antihumanism, 40, 49
antinaturalism, deconstruction as, 30
aporetic experience, 121–22
aporetic schematization, deconstructive
disposition and, 82
aporetic temporality, 90–94
Aporias (Derrida), 123
Aristotle, phronesis and, 68

251

David Wood is W. Alton Jones Professor of Philosophy at Vanderbilt University. His most recent book is *Deep Time, Dark Times: On Being Geologically Human.*

gROUNDWORKS|

ECOLOGICAL ISSUES IN PHILOSOPHY AND THEOLOGY

Forrest Clingerman and Brian Treanor, *series editors*

Interpreting Nature: The Emerging Field of Environmental Hermeneutics
Forrest Clingerman, Brian Treanor, Martin Drenthen,
and David Utsler, eds

*The Noetics of Nature: Environmental Philosophy
and the Holy Beauty of the Visible*
Bruce V. Foltz

*Environmental Aesthetics: Crossing Divides
and Breaking Ground*
Martin Drenthen and Jozef Keulartz, eds

The Logos of the Living World: Merleau-Ponty, Animals, and Language
Louise Westling

Being-in-Creation: Human Responsibility in an Endangered World
Brian Treanor, Bruce Ellis Benson, and Norman Wirzba, eds

*Wilderness in America: Philosophical Writings. Edited by
David W. Rodick*
Henry Bugbee

Eco-Deconstruction: Derrida and Environmental Philosophy
Matthias Fritsch, Philippe Lynes, and David Wood, eds

Animality: A Theological Reconsideration
Eric Daryl Meyer

Reoccupy Earth: Notes toward an Other Beginning
David Wood

Printed and bound by CPI Group (UK) Ltd, Croydon, CR0 4YY

27/10/2024

14580327-0002